Frank D. Petruzella

LogixPro PLC Lab Manual for use with

Programmable Logic Controllers

Fifth Edition

Mc
Graw
Hill
Education

LogixPro PLC Lab Manual for use with
PROGRAMMABLE LOGIC CONTROLLERS, FIFTH EDITION
Frank D. Petruzella

Some ancillaries, including electronic and print components, may not be available to customers outside the United States.

This book is printed on acid-free paper.

2 3 4 5 6 7 8 9 LOV 24 23 22 21 20

ISBN   978-1-259-68084-7
MHID   1-259-68084-3

mheducation.com/highered

Contents

Preface

This lab manual contains more than 250 programming assignments designed using the LogixPro simulation software. LogixPro essentially converts a personal computer into a PLC and allows you to write ladder logic programs and verify their real-world operation. LogixPro is a great tool for learning the fundamentals of Allen Bradley's RSLogix ladder software. It is ideally suited for students who don't have access to the actual PLC hardware and its programming software and for use in on-line courses. The simulated lab exercises parallel the first thirteen chapters in *Programmable Logic Controllers,* 5/e. The purpose of these exercises is to provide students with the opportunity to familiarize themselves with the many different features associated with PLCs. Programming assignments involve timers, counters, program control, data manipulation, math, sequencer, and shift register instructions. LogixPro also simulates a series of real-world equipment that the PLC will control. The real-world equipment is graphically animated on the computer screen. Answers to all the LogixPro programming assignments can be found in files on the *Instructor Resource section of Connect for Programmable Logic Controllers,* 5e. LogixPro is not included with this manual, so visit www.thelearningpit.com to purchase and download the software.

      I would like to thank Bill Simpson, creator of the LogixPro simulation software, for his involvement in the production of this text over the years. I hope you will find this simulation lab manual to be a helpful aid in understanding the operation of programmable controllers.

Frank D. Petruzella

About the Author

Frank D. Petruzella has extensive practical experience in the electrical control field, as well as many years of experience teaching and authoring textbooks. Before becoming a full-time educator, he was employed as an apprentice and electrician in areas of electrical installation and maintenance. He holds a Master of Science degree from Niagara University, a Bachelor of Science degree from the State University of New York College–Buffalo, as well as diplomas in Electrical Power and Electronics from the Erie County Technical Institute.

Getting Started with RSLogix & LogixPro

LogixPro allows you to practice and develop your RSLogix programming skills where and when you want. It replaces the PLC, ladder rung editor, and all the electrical components that have, until now, been required to learn RSLogix. It doesn't, however, replace instructors, texts, tutorials, or PLC documentation manuals, which are so essential when learning about PLCs and RSLogix. Think of LogixPro as a word processor, which will allow you to practice your literary skills *after* you are familiar with the meaning of words and how they are spelled.

The most commonly used elements of LogixPro are displayed below. The **Edit Panel** provides easy access to all the RSLogix instructions, and they may be simply dragged and dropped into your program.

Once your program is ready for testing, clicking on the "Toggle Button" of the Edit Panel will bring the **PLC Panel** into view. From the PLC Panel you can download your program to the "PLC" and then place it into the "RUN" mode. This will initiate the scanning of your program and the I/O of your chosen simulation.

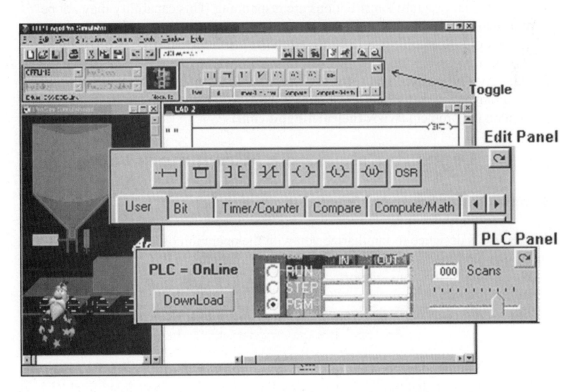

Editing Your Program

If you're familiar with Windows and how to use a mouse, then you are going to find editing a breeze. Both Instructions and Rungs are selected simply by clicking on them with the left mouse button. Deleting is then just a matter of hitting the DEL, or DELETE, key on your keyboard.

Double-clicking (two quick clicks) with the left mouse button allows you to edit an Instruction's address, while right clicking (with the right mouse button) displays a pop-up menu of related editing commands.

By clicking on an Instruction or Rung with the left mouse button and keeping it held down, you will be able to drag it wherever you please. Let go of it on any of the tiny locating boxes that you will see, and the Instruction or Rung will cling to its new home.

Debugging Your Program

If you take a look at the PLC Panel you'll notice an adjustable Speed Control. This is not a component of normal PLCs, but is provided with LogixPro so that you may adjust the speed of the simulations to suit your particular computer.

When the simulation is slowed, so is the PLC scanning. You can use this to good effect when trying to debug your program. Set the scan slow enough, and you can easily monitor how your program's instructions are responding. This capability may not be typical of real PLCs, but for training purposes, you will find that it is an invaluable debugging tool

If you run into problems with the exercises or LogixPro itself, remember that this is still relatively new material and software. Don't waste too much time trying to figure out some program action that seems amiss. Just e-mail us, and we'll do our best either to fix or to explain the problem (**www.thelearningpit.com**).

LogixPro Relay Logic Introductory Lab

RSLogix Relay Logic Instructions

This exercise is designed to familiarize you with the operation of LogixPro and to step you through the process of creating, editing, and testing simple PLC programs utilizing the Relay Logic Instructions supported by RSLogix.

From the Simulations menu at the top of the LogixPro window, select the I/O Simulator and ensure that the User Instruction Bar shown above is visible.

```
000  ├──────────────────────────────────────────────────( END )─┤
```

The program editing window should contain a single rung. This is the End of Program rung and is always the last rung in any program. If this is the only rung visible, then your program is currently empty.

If your program is not empty, then click on the File menu entry at the top of the window and select "New" from the drop-down list. A dialog box will appear asking you to select a Processor Type. Just click on "OK" to accept the default TLP LogixPro selection. Now maximize the ProSim-II Simulations window.

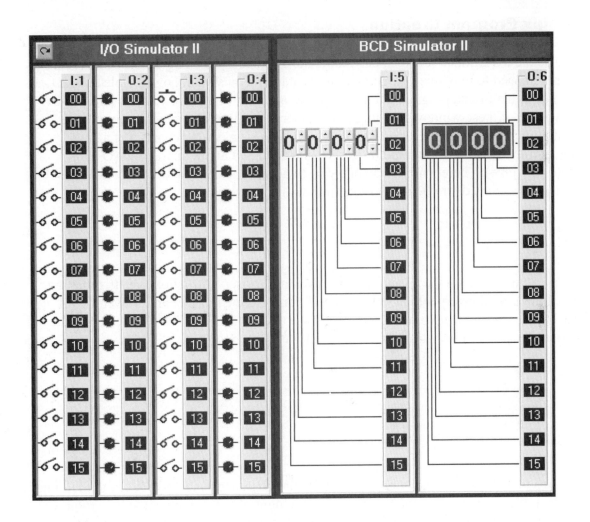

The I/O Simulator

The simulator screen shown above should now be in view. For this exercise, we will be using the I/O Simulator section, which consists of 32 switches and 32 lights. Two groups of 16 toggle switches are shown connected to two input cards of our simulated PLC. Likewise two groups of 16 lights are connected to two output cards of our PLC. The two input cards are addressed as "I:1" and "I:3," while the output cards are addressed "O:2" and "O:4."

Use your mouse to left click on the various switches, and note the change in the status color of the terminal that the switch is connected to. Move your mouse slowly over a switch, and the mouse cursor should change to a hand symbol, indicating that clicking at this location can alter the state of switch. When you pass the mouse over a switch, a "Tooltip" text box appears and informs you to "Right click to toggle switch type." Click your right mouse button on a switch, and note how the switch type may be readily changed.

RSLogix Program Creation

Collapse the I/O Simulator screen back to its normal size by clicking on the same (center) button you used to maximize the Simulator's window. You should now be able to see both the simulator and program editing windows again. If you wish, you can adjust the relative size of these windows by dragging the bar that divides them with your mouse. I want you to now enter the following single rung program, which consists of a single Input instruction (XIC—Examine If Closed) and a single Output instruction (OTE— Output Energize). There's more than one way to accomplish this task, but for now I will outline what I consider to be the most commonly used approach.

First click on the "New Rung" button in the User Instruction Bar. It's the first button on the very left end of the bar. If you hold the mouse pointer over any of these buttons for a second or two, you should see a short "Tooltip," which describes the function or name of the instruction that the button represents.

You should now see a new rung added to your program as shown above, and the rung number at the left side of the new rung should be highlighted. Note that the new rung was inserted above the existing End of Program (END) rung. Alternatively you could have dragged (with the left mouse button held down) the New Rung button into the program editing window and dropped it onto one of the locating boxes that would have appeared.

Make sure the rung number 000 is highlighted by left clicking on it. Now left click on the XIC instruction, and it will be added to the right of your highlighted selection. Note that the new XIC instruction is now selected (highlighted). Once again, you could have alternatively dragged and dropped the instruction into the program editing window.

If you accidentally add an instruction, which you wish to remove, just left click on the instruction to select it, and then press the "Del," or "Delete," key on your keyboard. Alternatively, you may right click on the instruction and then select "Cut" from the drop-down menu that appears.

Left click on the OTE output instruction, and it will be added to the right of your current selection.

```
          |?|                                                  ?
000    ┤ ├─────────────────────────────────────────────⟨ ⟩
```

Double-click (two quick left mouse button clicks) on the XIC instruction, and a text box should appear which will allow you to enter the address (I:1/0) of the switch we wish to monitor. Use the Backspace key to get rid of the "?" currently in the text box. Once you type in the address, press the Enter key on your keyboard or click anywhere else other than the text box, and the box should close.

Right click on the XIC instruction, and select "Edit Symbol" from the drop-down menu that appears. Another text box will appear where you can type in a name (Switch-0) to associate with this address. As before, pressing Enter or clicking anywhere else will close the box.

```
          I:1/0                                            0:2/0
000    ┤ ├─────────────────────────────────────────────⟨ ⟩
         Switch-0                                         Lamp-0
```

Enter the address and symbol for the OTE instruction, and your first RSLogix program will now be complete. Before continuing, however, you should double check that the addresses of your instructions are correct.

Testing your Program

It's now time to "Download" your program to the PLC. First click on the "Toggle" button at the top right corner of the Edit Panel, which will bring the PLC Panel into view.

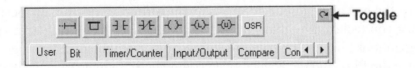 ← **Toggle**

Click on the "Download" button to initiate the downloading of your program to the PLC. Once complete, you should click inside the "RUN" option selection circle to start the PLC scanning.

Enlarge the ProSim Simulations window so that you can see both the switches and lamps, by dragging the bar that separates the Simulations and Program Editing windows to the right with your mouse. Now click on switch I:1/00 in the simulator, and if all is well, lamp O:2/00 should illuminate.

Toggle the Switch On and Off a number of times, and note the change in value indicated in the PLC Panel's status boxes that are being updated constantly as the PLC scans. Try placing the PLC back into the "PGM" mode, and then toggle the simulator's switch a few times, and note the result. Place the PLC back into the "RUN" mode, and the scan should resume.

We are usually told to think of the XIC instruction as an electrical contact that allows electrical flow to pass when an external switch is closed. We are then told that the OTE will energize if the flow is allowed to get through to it. In fact the XIC is a conditional instruction, which tests any bit that we address for truth or a 1.

Editing your Program

Click on the "Toggle" button of the PLC Panel, which will put the PLC into the PGM mode and bring the Edit Panel back into view.

Now add a second rung to your program as shown below. This time instead of entering the addresses as you did before, try dragging the appropriate address which is displayed in the I/O Simulator and dropping it onto the instruction.

Note that the XIO instruction, which tests for zero or false, has its address highlighted in yellow. This indicates that the instruction is true, which in the case of an XIO means that the bit addressed is currently a zero or false.

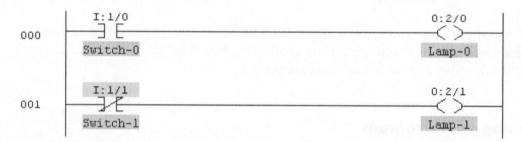

Try moving instructions from rung to rung by holding the left mouse button down while over an instruction, and then while keeping the mouse button down, move the mouse (and instruction) to a new location. Try doing the same with complete rungs by dragging the box at the left end of the rung and dropping it in a new location.

Once you feel comfortable with dragging and dropping, ensure that your program once again looks like the one pictured above. Now download your program to the PLC, and place the PLC into the RUN mode. Toggle both Switch-0 and Switch-1 on and off a number of times, and observe the effects this has on the lamps. Ensure that you are satisfied with the operation of your program before proceeding further.

CHAPTER **1**

Programmable Logic Controllers (PLCs): An Overview

LogixPro Programming Assignments

1-1(a) Write a documented program for the relay schematic shown below. Use the I/O Simulator screen and the following addresses to simulate the program:

Pressure switch _ I:1/0
Temperature switch _ I:1/1
Manual pushbutton _ I:1/2
Motor starter coil _ O:2/0

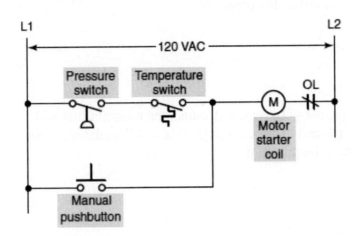

1-1(b) Modify the original program to operate according to the relay schematic shown below.

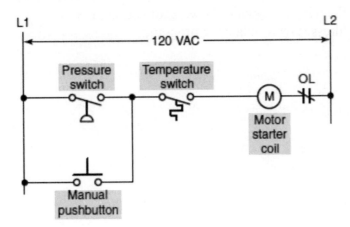

1-1(c) Modify the original program to operate so that the manual pushbutton, pressure switch, and temperature switch all must be closed to energize the motor starter.

1-1(d) Modify the original program to operate so that closing the manual pushbutton, pressure switch, or the temperature switch will energize the motor starter.

1-2 Given two single pole switches, write a documented program that will turn a lamp on when both switch A and switch B are closed. Use the I/O Simulator screen and the following addresses to simulate the program:

Switch A _ I:1/0
Switch B _ I:1/1
Lamp _ O:2/0

1-3 Given two normally open pushbuttons, write a documented program that will turn a lamp on when either pushbutton A or pushbutton B is closed. Use the I/O Simulator screen and the following addresses to simulate the program:

Pushbutton A _ I:1/0
Pushbutton B _ I:1/1
Lamp _ O:2/0

1-4 Given four single pole switches (A-B-C-D), write a documented program that will turn on a lamp if switches A and B or C and D are closed. Use the I/O Simulator screen and the following addresses to simulate the program:

Switch A _ I:1/0
Switch B _ I:1/1
Switch C _ I:1/2
Switch D _ I:1/3
Lamp _ O:2/0

1-5 Write a documented program for the relay schematic shown below. Use the I/O Simulator screen and the following addresses to simulate the program:

S1 _ I:1/0
LS1 _ I:1/1
LS2 _ I:1/2
L1 _ O:2/0

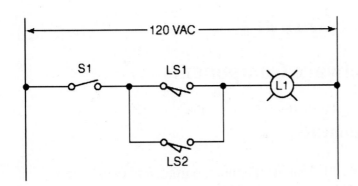

1-6 Write a documented program for the relay schematic shown below.

Use the I/O Simulator screen and the following addresses to simulate the program:

PB1 _ I:1/0
S1 _ I:1/1
S2 _ I:1/2
S3 _ I:1/3
PS1 _ I:1/4
TS1 _ I:1/5
L1 _ O:2/0

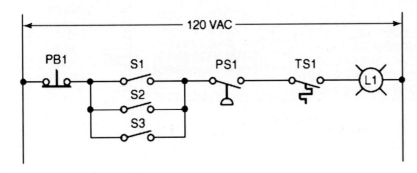

CHAPTER **2**

PLC Hardware Components

LogixPro Programming Assignments

2-1 Write a documented program that will simulate the input, output, and data table conditions for the PLC circuit shown below. Use the I/O Simulator screen and the following addresses to simulate the program:

NC pushbutton _ I:1/1
NO limit switch _ I:1/8
L1 _ O:2/2
L2 _ O:2/7

Observe the status of the bits stored in the input and output image tables.

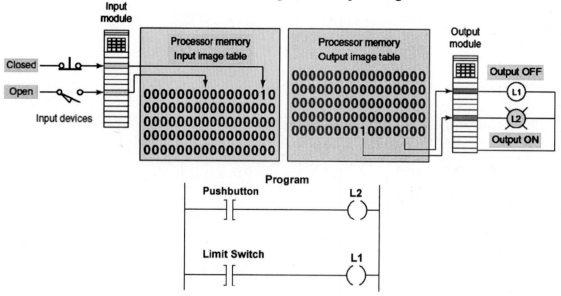

2-2 Write a program that will produce the following binary numbers in input word I:1 and output word O:2 of the data table. Write one rung for each output bit, and select either an XIC or XIO of the corresponding input bit to produce the corresponding output.

DATA TABLE

15	14	13	12	11	10	9	8	7	6	5	4	3	2	1	0

(Determined by the ON/OFF state of the inputs)

I:1	1	0	1	1	0	0	1	1	0	0	0	1	1	1	0	1

(Determined by the program)

O:2	0	0	0	0	0	0	0	0	0	0	0	0	1	1	1	1

Use the I/O Simulator screen to simulate the program. Use the radix function to determine decimal, hex-BCD, and octal values for binary numbers of input word I:1 and output word O:2.

CHAPTER **3**

Number Systems and Codes

LogixPro Programming Assignments

3-1 Operate a simulation for the single-digit BCD thumb-wheel switch interfaced to the PLC circuit shown below. Use the BCD simulator screen and the following addresses to simulate the program:

1s _ I:1/00
2s _ I:1/01
4s _ I:1/02
8s _ I:1/03

Record the binary input values (0 or 1) found in the Input Table for decimal thumb-wheel switch settings 0 through 9.

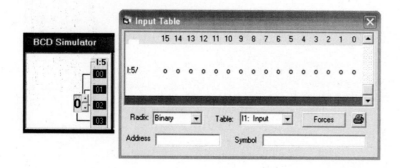

3-2 Operate a simulation for the double-digit BCD thumb-wheel switch interfaced to the PLC circuit shown below. Use the BCD simulator screen and the following addresses to simulate the program:

1s _ I:1/00
2s _ I:1/01
4s _ I:1/02
8s _ I:1/03
1s _ I:1/04
2s _ I:1/05
4s _ I:1/06
8s _ I:1/07

Record the binary input values (0 or 1) found in the Input Table for decimal thumb-wheel switch settings 0 through 20.

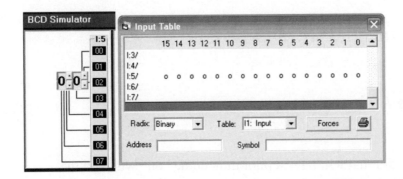

3-3 Simulate the monitoring of the setting of a thumb-wheel switch program shown below. Use the BCD simulator screen and the addresses shown.

- The Convert from BCD Instruction (**FRD**) is used to convert 16-bit integers into BCD (Binary Coded Decimal) values.
- The Convert to BCD Instruction (**TOD**) is used to convert BCD values into integers.

Operate the program and record the binary input value for B3:0 for a thumb-wheel switch setting of 20.

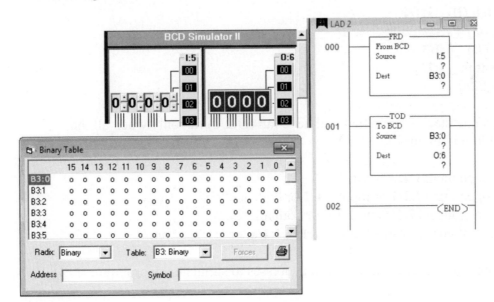

CHAPTER 4

Fundamentals of Logic

LogixPro Programming Assignments

4-1 Write a documented program that will have a pilot light come "on" when all the following circuit requirements are met:

➢ Both normally open circuit pressure switches (A & B) must be closed.
➢ One of two normally open circuit limit switches (C & D) must be closed.
➢ The normally open reset pushbutton must not be closed.

Use the I/O Simulator screen and the following addresses to simulate the program:

Pressure switch (A) _ I:1/0
Pressure switch (B) _ I:1/1
Limit switch (C) _ I:1/2
Limit switch (D) _ I:1/3
Reset pushbutton _ I:1/4
Pilot light _ O:2/0

4-2 Write a documented program that will have fan **(O:2/0)** run only when all the following conditions are met:

➢ Normally open input **I:1/0** is "off."
➢ Normally open input **I:1/1** or input **I:1/2** or both are "on."
➢ Normally open inputs **I:1/3** and **I:1/4** or both are "on."
➢ One or more of normally open inputs **I:1/5, I:1/6,** or **I:1/7** is "on."

Use the I/O Simulator screen and the given addresses to simulate the program.

4-3 Write a documented program for the relay schematic shown. Use the I/O Simulator screen and the following addresses to simulate the program:

LS1 _ I:1/0
LS2 _ I:1/1
SOL _ O:2/0

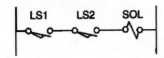

Relay schematic

4-4 Write a documented program for the relay schematic shown. Use the I/O Simulator screen and the following addresses to simulate the program:

LS1 _ I:1/0
LS2 _ I:1/1
SOL _ O:2/0

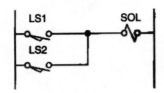

Relay schematic

4-5 Write a documented program for the relay schematic shown. Use the I/O Simulator screen and the following addresses to simulate the program:

LS1 _ I:1/0
LS2 _ I:1/1
PS _ I:1/2
PL _ O:2/0

Relay Schematic

4-6 Write a documented program for the relay schematic shown. Use the I/O Simulator screen and the following addresses to simulate the program:

LS1 _ I:1/0
LS2 _ I:1/1
FS1 _ I:1/2
FS2 _ I:1/3
PL _ O:2/0

Relay Schematic

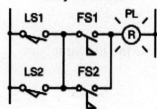

4-7 Write a documented program for the relay schematic shown. Use the I/O Simulator screen and the following addresses to simulate the program:

LS1 _ I:1/0
LS2 _ I:1/1
LS3 _ I:1/2
Horn _ O:2/0

Relay schematic

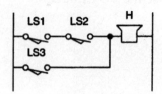

4-8 Write a documented program for the relay schematic shown. Use the I/O Simulator screen and the following addresses to simulate the program:

LS1 _ I:1/0
LS2 _ I:1/1
LS3 _ I:1/2
LS4 _ I:1/3
PL _ O:2/0

Relay schematic

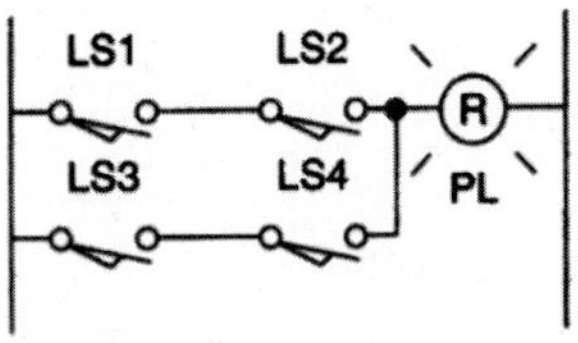

4-9 Write a documented program for the relay schematic shown. Use the I/O Simulator screen and the following addresses to simulate the program:

LS1 _ I:1/0
PB _ I:1/1
SOL _ O:2/0

Relay Schematic

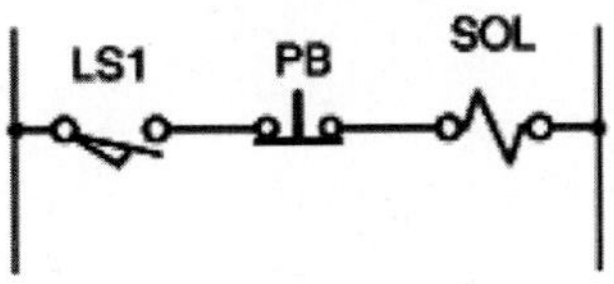

4-10 Write a documented program, which will express the Boolean equation

Y = (A+B) CD

as a ladder logic rung. Use the I/O Simulator screen and the following addresses to simulate the program:

A _ I:1/0
B _ I:1/1
C _ I:1/2
D _ I:1/3
Y _ O:2/0

4-11 Write a documented program, which will express the Boolean equation

$$Y = (A\overline{B}C) + \overline{D} + E$$

as a ladder logic rung. Use the I/O Simulator screen and the following addresses to simulate the program:

A _ I:1/0
B _ I:1/1
C _ I:1/2
D _ I:1/3
E _ I:1/4
Y _ O:2/0

4-12 Write a documented program, which will express the Boolean equation

$$Y = [(\overline{A} + \overline{B})C] + DE$$

as a ladder logic rung. Use the I/O Simulator screen and the following addresses to simulate the program:

A _ I:1/0
B _ I:1/1
C _ I:1/2
D _ I:1/3
E _ I:1/4
Y _ O:2/0

4-13 Write a documented program, which will express the Boolean equation

$$Y = (\overline{A}B\overline{C}) + (D\overline{E}F)$$

as a ladder logic rung. Use the I/O Simulator screen and the following addresses to simulate the program:

A _ I:1/0
B _ I:1/1
C _ I:1/2
D _ I:1/3
E _ I:1/4
F _ I:1/5
Y _ O:2/0

12

4-14 Write a documented program which will simulate the operation of the "XOR" function $(\overline{A}B + A\overline{B}) = Y$.

Use the I/O Simulator screen and the following addresses to simulate the program:

A _ I:1/0
B _ I:1/1
Y _ O:2/0

4-15 A conveyor will run when any one of two inputs is "on." It will stop when any one of two other inputs is "on." Write a documented program which will simulate this operation. Use the I/O Simulator screen and the following addresses to simulate the program:

ON-Input _ I:1/0
ON-Input _ I:1/1
OFF-Input _ I:1/2
OFF-Input _ I:1/3
Conveyor _ O:2/0

4-16 Write a documented program, which will simulate the gate array logic shown as a single ladder logic rung. Use the I/O Simulator screen and the following addresses to simulate the program:

A _ I:1/0
B _ I:1/1
C _ I:1/2
D _ I:1/3
Y _ O:2/0

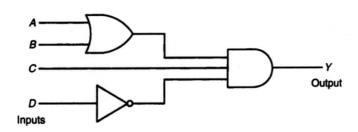

4-17 Write a documented program, which will simulate the gate array logic shown as a single ladder logic rung. Use the I/O Simulator screen and the following addresses to simulate the program:

A _ I:1/0
B _ I:1/1
Y _ O:2/0

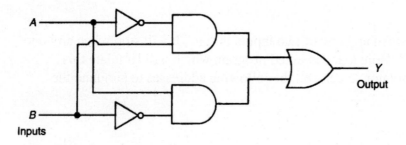

4-18 Write a documented program, which will simulate the gate array logic shown as a single ladder logic rung. Use the I/O Simulator screen and the following addresses to simulate the program:

A _ I:1/0
B _ I:1/1
C _ I:1/2
D _ I:1/3
M _ O:2/0

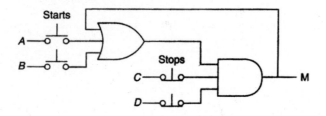

4-19 Write a documented program, which will simulate the gate array logic shown as a single ladder logic rung. Use the I/O Simulator screen and the following addresses to simulate the program:

A _ I:1/0
B _ I:1/1
C _ I:1/2
D _ I:1/3
Y _ O:2/0

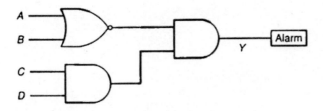

4-20 Write a documented program, which will simulate the gate array logic shown as a single ladder logic rung. Use the I/O Simulator screen and the following addresses to simulate the program:

A _ I:1/0
B _ I:1/1
C _ I:1/2
Y _ O:2/0

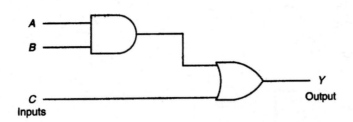

4-21 Write a documented program, which will simulate the gate array logic shown as a single ladder logic rung. Use the I/O Simulator screen and the following addresses to simulate the program:

A _ I:1/0
B _ I:1/1
C _ I:1/2
D _ I:1/3
Y _ O:2/0

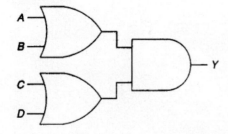

4-22 Write a documented program, which will simulate the gate array logic shown as a single ladder logic rung. Use the I/O Simulator screen and the following addresses to simulate the program:

A _ I:1/0
B _ I:1/1
C _ I:1/2
D _ I:1/3
E _ I:1/4
Y _ O:2/0

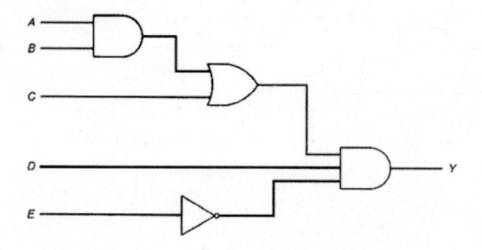

4-23 Write a documented program, which will simulate the gate array logic shown as a single ladder logic rung. Use the I/O Simulator screen and the following addresses to simulate the program:

A _ I:1/0
B _ I:1/1
C _ I:1/2
D _ I:1/3
E _ I:1/4
F _ I:1/5
Y _ O:2/0

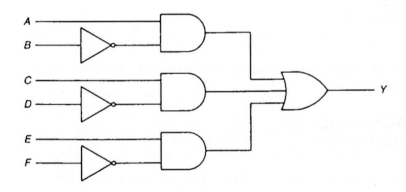

4-24 Simulate the AND instruction program shown below. Use the I/O Simulator screen and the addresses given. Set the value of:

Source A (B3:5) to 0000000010101010
Source B (B3:7) to 0000000011101011

Operate the circuit and record the resultant binary value at the **Destination (B3:10)**.

```
   Input A    ┌─AND──────────────────┐
 ──┤ ├─────── │ BITWISE AND          │
    I:1/0     │ Source A       B3:5  │
              │ 0000000010101010     │
              │ Source B       B3:7  │
              │ 0000000011101011     │
              │ Destination    B3:10 │
              │ ??????????????????   │
              └──────────────────────┘
```

4-25 Simulate the OR instruction program shown below. Use the I/O Simulator screen and the addresses given. Set the value of:

Source A (B3:1) to 1100110011001100
Source B (B3:2) to 1111111100000000

Operate the circuit and record the resultant binary value at the **Destination (B3:20)**.

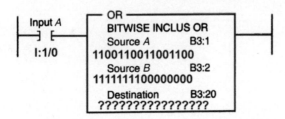

4-26 Simulate the XOR instruction program shown below. Use the I/O Simulator screen and the addresses given. Set the value of:

Source A (I:1.0) to 0000000010101010
Source B (I:3.0) to 0000000011101011

Operate the circuit and record the resultant binary value at the **Destination (O:4.0)**.

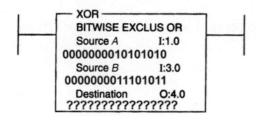

18

4-27 Simulate the NOT instruction program shown below. Use the I/O Simulator screen and the addresses given. Set the value of:

Source (B3:9) to 0000000010101010

Operate the circuit and record the resultant binary value at the **Destination (B3:10)**.

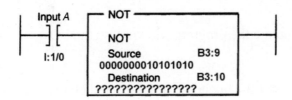

CHAPTER **5**

Basics of PLC Programming

LogixPro Programming Assignments

5-1 Write a documented program that will simulate the program shown. Use the I/O Simulator screen and the addresses shown.

With input I:3/12 **not actuated,** record the state of the bit stored at I:3/12 and O:4/6.

With input I:3/12 **actuated,** record the state of the bit stored at I:3/12 and O:4/6.

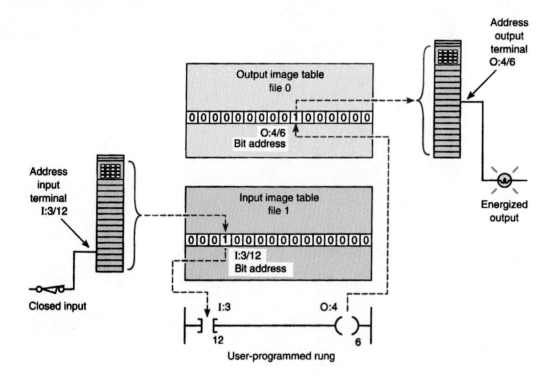

5-2 Write a documented program that will implement the ladder logic shown. Write the Boolean equation for this program. Use the I/O Simulator screen and the following addresses to simulate the program:

A _ I:1/0
B _ I:1/1
C _ I:1/2
D _ I:1/3
E _ I:1/4
Y _ O:2/0

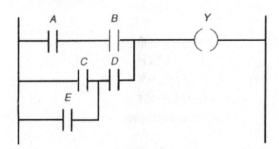

5-3 Write a documented program that will implement the ladder logic shown. Write the Boolean equation for this program. Use the I/O Simulator screen and the following addresses to simulate the program:

A _ I:1/0
B _ I:1/1
C _ I:1/2
D _ I:1/3
Y _ O:2/0

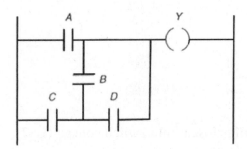

5-4 Write a documented program that will implement the ladder logic shown. Write the Boolean equation for this program. Use the I/O Simulator screen and the following addresses to simulate the program:

A _ I:1/0
B _ I:1/1
C _ I:1/2
D _ I:1/3
E _ I:1/4
Y _ O:2/0

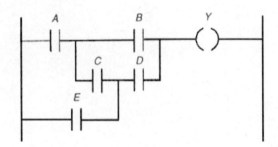

5-5(a) Write a documented program that will implement the hardwired limit switch circuit shown using a **single normally open limit switch contact.** Use the I/O Simulator screen and the following addresses to simulate the program:

LS1 _ I:1/0
SOL A _ O:2/0
SOL B _ O:2/1

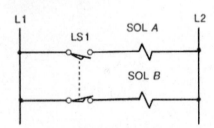

5-5(b) Modify the program to allow a single **normally closed** limit switch contact (**LS2 _ I:1/1**) to be used as the input to implement the same logic.

5-6 Write a documented program that will implement the hardwired relay control circuit shown **without** the use of control relays CR1, CR2, and CR3. Use the I/O Simulator screen and the following addresses to simulate the program:

PB1 _ I:1/2	PB3 _ I:1/5
LS1 _ I:1/0	PL1 _ O:2/0
PS1 _ I:1/1	SOL A _ O:2/1
SS1 _ I:1/6	SOL B _ O:2/2
LS2 _ I:1/7	SOL C _ O:2/3
LS3 _ I:1/4	PL2 _ O:2/4
PB2 _ I:1/3	

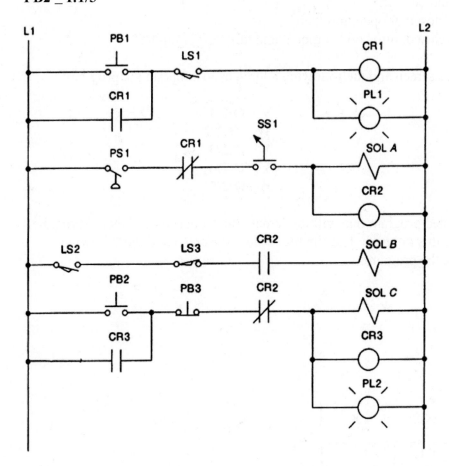

5-7 Write a documented program that will implement all of the following tasks:

➢ Turn on a light when a conveyor motor (O:2/0) is running in reverse. The input field device (I:1/0) is a set of contacts on the conveyor reversing starter that close when the motor is running in the forward direction and open when it is running in reverse.

➢ When a pushbutton is pressed, it operates a solenoid (O:2/1). The input field device is a normally open pushbutton (I:1/3).

> Stop a motor (O:2/2) from running when a pushbutton is pressed. The input field device is a normally closed pushbutton (I:1/4).
> When a limit switch is closed, it triggers instruction (O:2/3) "on." The input field device is a limit switch (I:1/5) that stores 1 in a data file bit when closed.

Use the I/O Simulator screen and the specified addresses to simulate the program.

5-8 Write a documented program that will implement all of the following tasks:

> When input A is closed, turn "on" and hold "on" outputs U and V until A opens.
> When input A is closed and either input B or C is open, turn "on" output W; otherwise, it should be "off."
> When input D is closed or open, turn "on" output X.
> When input D is closed, turn "on" output Y and turn "off" output Z.

Use the I/O Simulator screen and the following addresses to simulate the program:

A _ I:1/0	V _ O:2/1
B _ I:1/1	W _ O:2/2
C _ I:1/2	X _ O:2/3
D _ I:1/3	Y _ O:2/4
U _ O:2/0	Z _ O:2/5

5-9 Write a documented program that will implement the ladder logic shown. Write the Boolean equation for this program. Use the I/O Simulator screen and the following addresses to simulate the program:

A _ I:1/0
B _ I:1/1
C _ I:1/2
D _ I:1/3
E _ I:1/4
Y _ O:2/0

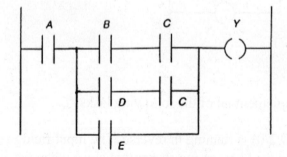

5-10 Write a documented program that will implement the ladder logic shown. Write the Boolean equation for this program. Use the I/O Simulator screen and the following addresses to simulate the program:

A _ I:1/0
B _ I:1/1
C _ I:1/2
D _ I:1/3
Y _ O:2/0

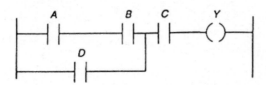

5-11 Write a documented program that will implement the ladder logic shown. Write the Boolean equation for this program. Use the I/O Simulator screen and the following addresses to simulate the program:

A _ I:1/0
B _ I:1/1
C _ I:1/2
D _ I:1/3
Y _ O:2/0

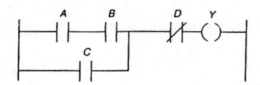

5-12(a) Write a documented program that will implement the start/stop ladder logic program shown. Use the I/O Simulator screen and the following addresses to simulate the program:

A (NO stop pushbutton) _ I:1/0
B (NO start pushbutton) _ I:1/1
Y _ O:2/0

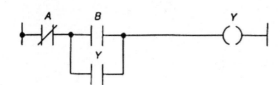

5-12(b) Modify the original program to operate in the same manner using a normally closed stop pushbutton in place of the normally open pushbutton.

5-13 Write a documented program that will implement the ladder logic shown. Use the I/O Simulator screen and the following addresses to simulate the program:

A _ I:1/0
B _ I:1/1
V _ O:2/0
W _ O:2/1
X _ O:2/2
Y _ O:2/3
Z _ O:2/4

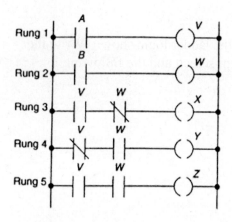

List what outputs will be energized for each of the following input conditions:

    I.   Rung 1 has logic continuity but Rung 2 does not.
   II. Rung 2 has logic continuity but Rung 1 does not.
  III. Both Rung 1 and Rung 2 have logic continuity.

5-14(a) Write a documented program that will implement the ladder logic shown for a light controlled from two positions using 2 SPST switches (in place of two 3-way switches). Use the I/O Simulator screen and the following addresses to simulate the program:

S1 _ I:1/0
S2 _ I:1/1
Light _ O:2/0

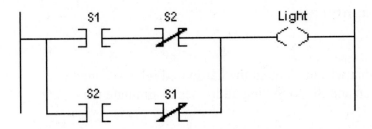

5-14(b) Add a third SPST switch to have the light controlled from three positions using three SPST switches (in place of two 3-way switches and one 4-way switch). Use the I/O Simulator screen and the following addresses to simulate the program:

S1 _ I:1/0
S2 _ I:1/1
S3 _ I:1/2
Internal Relay Bit _ B3:1/0
Light _ O:2/0

CHAPTER **6**

Developing Fundamental PLC Wiring Diagrams and Ladder Logic Programs

LogixPro Programming Assignments

6-1 Write a documented program that will implement the hardwired relay schematic shown. Use the I/O Simulator screen and the following addresses to simulate the program:

PB1 _ I:1/0
Horn _ O:2/0
Sol _ O:2/1
M _ O:2/2
PL _ O:2/3

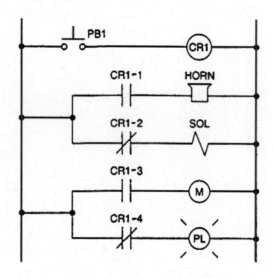

6-2 Write a documented program that will implement the hardwired relay schematic shown using only the single set of normally open contacts of the pressure switch. Use the I/O Simulator screen and the following addresses to simulate the program:

Pressure Switch _ I:1/0
L _ O:2/0
H _ O:2/1

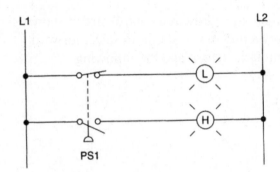

6-3(a) Write a documented program that will implement the standard hardwired start/stop motor control circuit shown. Use the Silo Simulator screen and the following addresses to simulate the program:

START _ I:1/0
STOP _ I:1/1
M _ O:2/0

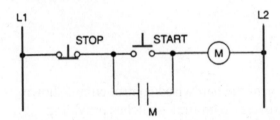

6-3(b) Add the necessary programming for a motor "RUN" light **(O:2/2)** and a motor "STANDBY" or "OFF" light **(O:2/3)**.

6-3(c) Add the necessary programming so that the motor will not operate unless the selector switch is at position "A" with its NO contact **I:1/5** closed.

6-4 Write a documented program that will allow a motor to be started and stopped from any three start/stop pushbutton stations. Use the I/O Simulator screen and the following addresses to simulate the program:

Stop PB1 (NC) _ I:1/0
Start PB2 (NO) _ I:1/1
Stop PB3 (NC) _ I:1/2
Start PB4 (NO) _ I:1/3
Stop PB5 (NC) _ I:1/4
Start PB6 (NO) _ I:1/5
Motor OL contacts (NC) _ I:1/6
Motor _ O:2/0

6-5 Write a documented program that will implement the hardwired control circuit shown that uses a selector switch in conjunction with a reversing motor starter to select forward or reverse operation of the motor. Use the I/O Simulator screen and the following addresses to simulate the program:

Stop _ I:1/0
Start _ I:1/1
Selector Switch _ I:1/2 (closed for forward)
OLs _ I:1/3
F _ O:2/0
R _ O:2/1

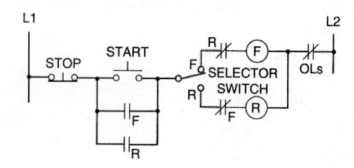

6-6 Write a documented program that will implement the hardwired control circuit shown for stopping and starting a motor in forward and reverse with limit switches providing over travel protection. Use the Door Simulator screen and the following addresses to simulate the program:

Forward PB _ I:1/0
Reverse PB _ I:1/1
Stop PB _ I:1/2
Forward Limit Switch _ I:1/3
Reverse Limit Switch _ I:1/4
F (Motor Up) _ O:2/0
R (Motor Down) _ O:2/1

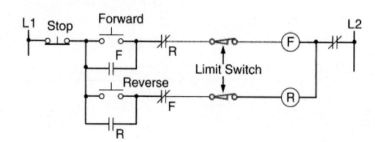

6-7 Write a documented program that will allow perform the following circuit functions:

> Turn "on" PL1 **(O:2/0)** when input S1 **(I:1/0)** is closed.
> Turn "on" PL2 **(O:2/1)** when input S2 **(I:1/1)** is closed.
> Electrically interlock the inputs so that the two lights cannot be both turned "on" at the same time.

Use the I/O Simulator screen and the designated addresses to simulate the program.

6-8 Write a documented program that will implement the manual/automatic relay control schematic shown. The sequence of operation is as follows:

> The pump is started by momentarily pressing the start button.
> With the selector switch in the manual position, the solenoid valve is energized at all times.
> With the selector switch in the automatic position, the solenoid valve is energized only when the motor is operating and the pressure switch is closed.

Use the I/O Simulator screen and the following addresses to simulate the program:

Stop _ I:1/0
Start _ I:1/1
Manual/Auto Switch _ I:1/2 (closed in the manual position)
PS1 _ I:1/3
M _ O:2/0
PL _ O:2/1
SV _ O:2/2

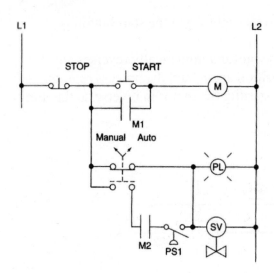

6-9 Write a documented program that will implement the forward/reverse motor starter, with electrical interlocks, relay schematic shown. Use the I/O Simulator screen and the following addresses to simulate the program:

Stop PB _ I:1/0
Fwd PB _ I:1/1
Rev PB _ I:1/2
F _ O:2/0
R _ O:2/1

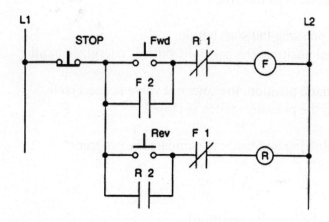

6-10 Write a documented program that will implement the hardwired reciprocating motion machine process control schematic shown. The sequence of operation is as follows:

➤ The workpiece starts on the left and moves to the right when the start button is momentarily actuated.
➤ When it reaches the rightmost limit (LS2), the motor automatically reverses and brings the workpiece back to the leftmost position again, and the process repeats.
➤ The reverse pushbutton provides a means of starting the motor in reverse so that limit switch LS1 can take over automatic control.

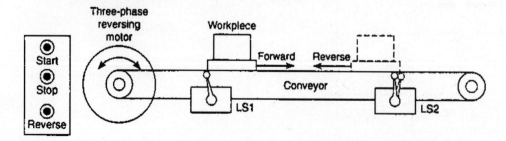

Use the Door Simulator screen and the following addresses to simulate the program:

Start PB _ I:1/0
Reverse PB _ I:1/1
Stop PB _ I:1/2
Limit Switch LS2 _ I:1/3
Limit Switch LS1 _ I:1/4
F (Motor Forward) _ O:2/0
R (Motor Reverse) _ O:2/1
Internal Relay _ B3:0/0

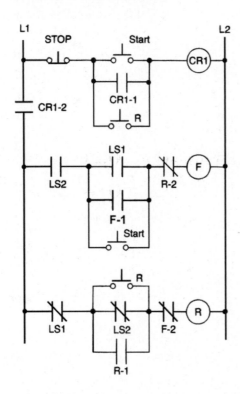

6-11(a) Write a documented program that will implement the following continuous filling operation:

➢ Start the conveyor when the Start button is momentarily pressed.
➢ Stop the conveyor when the Stop button is momentarily pressed.
➢ Energize the Run status light when the process is operating.
➢ Energize the Standby status light when the process is stopped.
➢ Stop the conveyor and energize the Standby light when the right edge of the box is first sensed by the photosensor.
➢ With the box in position and the conveyor stopped, open the solenoid valve and allow the box to fill. Filling should stop when the Level sensor goes true.
➢ Energize the Full light when box is full. The Full light should remain energized until the box is moved clear of the photosensor.

Use the Silo Simulator screen and the following addresses to simulate the program:

Start (NO contact) _ I:1/0
Stop (NC contact) _ I:1/1
Photo Switch (NO contact) _ I:1/3
Level Switch (NO contact) _ I:1/4
Conveyor Motor _ O:2/0
Solenoid _ O:2/1
Run Light _ O:2/2
Standby Light _ O:2/3
Full Light _ O:2/4

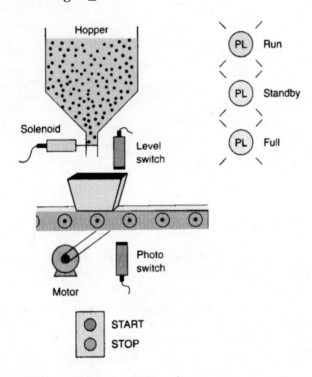

6-11(b) Modify the original program so that the panel mounted selector switch can be used to select one of the following three operating modes:

➢ When the selector switch is in position "A," the process shall be virtually turned off with no outputs being capable of being energized.
➢ When the sector switch is in position "B," the process shall operate as the continuous filling program of 6-11(a).
➢ When the sector switch is in position "C," the process shall operate in such a manner that the boxes will simply move down the conveyor continuously and bypass the fill operation.

6-12 Write a documented program that will cause output pilot light PL to be "on" when selector switch SS is closed, pushbutton PB is open, and limit switch LS is open. Use the I/O Simulator screen and the following addresses to simulate the program:

Selector Switch _ I:1/0
Pushbutton _ I:1/1
Limit Switch _ I:1/2
Pilot Light _ O:2/0

6-13 Write a documented program that will cause a solenoid, SOL, to be energized when limit switch LS is closed and pressure switch PS is open. Use the I/O Simulator screen and the following addresses to simulate the program:

Limit Switch _ I:1/0
Pressure Switch _ I:1/1
SOL _ O:2/0

6-14(a) Implement the Latch/Unlatch program shown using the I/O Simulator screen and the following addresses:

Input A _ I:1/0
Input B _ I:1/1
Output C _ O:2/0

Operate the program normally. Operate the program with both inputs A and B closed. Make note of the status of the PL ("on" or "off" at all times). With reference to the way the controller executes the program, explain why the light appears to be "on" or "off" at all times.

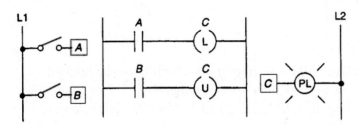

6-14(b) Reprogram the circuit as shown. Operate the program normally. Operate the program with both inputs A and B closed. Make note of the status of the PL ("on" or "off" at all times). With reference to the way the controller executes the program, explain why the light behaves differently with both switches closed than the above circuit.

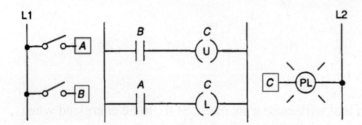

6-15 Program two start/stop motor control circuits that can achieve the desired result of the relay schematic shown. One design is to be seal-in type, start/stop control, and the other is to be a latch/unlatch type start/stop control. Explain when you would use the one circuit over the other. Use the I/O Simulator screen and the following addresses to simulate the program:

Seal-in { **Start PB _ I:1/0**
Circuit { **Stop PB _ I:1/1**
        { **M _ O:2/0**

Latch/Unlatch { **Start PB _ I:1/2**
Circuit Stop { **PB _ I:1/3**
           { **M _ O:2/1**

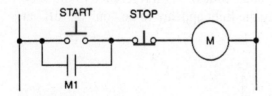

6-16 Write a documented program that will cause output pilot light PL to be latched when Switch 1 is momentarily closed, and unlatched when either Switch 2 or Switch 3 is momentarily closed. Also do not allow the unlatch to go TRUE when the latch rung is TRUE, nor allow the latch rung to go TRUE when the unlatch rung is TRUE. Use the I/O Simulator screen and the following addresses to simulate the program:

Switch 1 _ I:1/0
Switch 2 _ I:1/1
Switch 3 _ I:1/2
PL _ O:2/0

6-17(a) Write a program that will implement the following manual vessel filling operation:

➤ Any time NO Start PB **(I:1/0)** is pressed, pump P1 **(O:2/1)** will be energized and fluid will flow into the vessel.
➤ Any time NC Stop PB **(I:1/1)** is pressed, pump P3 **(O:2/3)** will be energized and fluid will be discharged from the vessel.
➤ Run pilot light **(O:2/5)** to come "on" anytime pump P1 is operating.
➤ Idle pilot light **(O:2/6)** to come "on" anytime pump P1 is not operating.
➤ Whenever the NO High Level sensor **(I:1/4)** is actuated, stop pump P1 if it is running or do not allow it to start if it is stopped.
➤ Whenever the NO Low Level sensor **(I:1/3)** is **not** actuated, stop pump P3 if it is running or do not allow it to start if it is stopped.
➤ Whenever the NO High Level sensor **(I:1/4)** is actuated, have the Full pilot light **(O:2/7)** come "on."

Use the Batch Simulator screen and the designated addresses to simulate the program.

6-17(b) Modify the original program to include the following features:

➤ A second pump P2 **(O:2/2)** that operates whenever P1 is operating.
➤ Turn the heater **(O:2/4)** "on" whenever the vessel is Full.
➤ Turn the Mixer motor **(O:2/0)** "on" whenever the vessel is at any level but Low.

6-18 Write a documented program that will control the level of water in a storage tank by turning a discharge pump on or off. The operation is as follows:

➤ With the control panel selector switch in position "A," when the level of the water in the tank reaches a high level, the discharge pump will **latch** "on" automatically and remain "on" until the water level reaches the low level.
➤ With the selector switch in position "B," the discharge pump will start automatically if the water in the tank is at any level except low and remain "on" until the water level reaches the low level.
➤ Use the control panel start/stop pushbutton station to operate the supply pump motor to change the level of water in the tank. Do not allow this input pump motor to operate if the high level has been reached.
➤ Have the RUN pilot light come "on" whenever the discharge pump is operating. Do not allow the discharge pump to operate if the supply pump motor is operating.

Use the Batch Simulator screen and the following addresses to simulate the program:

Start Fill PB (NO) _ I:1/0
Stop Fill PB (NC) _ I:1/1
Low Level Sensor _ I:1/3
High Level Sensor _ I:1/4
Position "A" _ I:1/9
Position "B" _ I:1/10
Latch/Unlatch Internal Relay _ B3:1/0
Supply Pump Motor _ O:2/1
Discharge Pump Motor _ O:2/3
Run PL _ O:2/5

6-19 Write a documented program that will perform the following switching tasks:

➢ When switch S1 is closed, lights L1, L2, and L3 come "on."
➢ When switch S2 is closed, lights L1 and L2 drop out, leaving light L3 "on."
➢ When switch S3 is closed (all three switches closed at this point), L1 will come "on," thus showing L1 and L3 "on."
➢ When switch S4 is closed, it turns "off" any of the lights that happen to be "on."

Use the I/O Simulator screen and the following addresses to simulate the program:

S1 _ I:1/0
S2 _ I:1/1
S3 _ I:1/2
S4 _ I:1/3
L1 _ O:2/0
L2 _ O:2/1
L3 _ O:2/2

6-20 There are four input sensors that control an alarm annunciator system in case an operational malfunction occurs. Design a documented program that operates the alarm system as follows:

➢ If any one sensor input is "closed," nothing happens.
➢ If any two inputs are "closed," a green pilot light goes "on."
➢ If any three inputs are "closed," a yellow pilot light goes "on."
➢ If all four inputs are "closed," a red pilot light goes "on."

Use the I/O Simulator screen and the following addresses to simulate the program:

S1 _ I:1/0
S2 _ I:1/1
S3 _ I:1/2
S4 _ I:1/3
Green PL _ O:2/0
Yellow PL _ O:2/1
Red PL _ O:2/2

6-21(a) The ladder logic program shown is a push-to-start/push-to-stop circuit. A single NO pushbutton **(I:1/0)** performs both the Start and Stop functions. The first time you press the pushbutton, instruction **B3:0/11** is latched, energizing light output **O:2/0**. The second time you press the pushbutton, instruction **B3:0/12** unlatches instruction **B3:0/11**, de-energizing output **O:2/0**. Instruction **B3:0/10** prevents interaction between **B3:0/12** and **B3:0/11.** Implement the program shown using the I/O Simulator screen and the designated addresses.

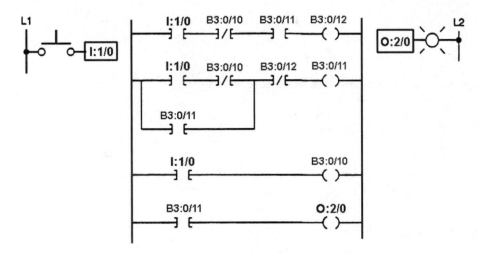

6-21(b) Modify the original program to control the light from four remote locations using two NO pushbuttons **(I:1/0 & I:1/1)** and two NC pushbuttons **(I:1/2 & I:1/3).** The program should result in the light changing state when any one of the four pushbuttons is pressed.

6-22 Write a program that will implement the conveyor relay schematic shown. Use the I/O Simulator screen and the following addresses to simulate the program:

Field Device	Logical Address	Symbolic Address
Start Button	I:3/0	PB1
Emergency Stop Button	I:3/1	PB2
Limit Switch	I:3/2	LS1
Motor Starter Coil	O:4/1	M
Red - Stop Pilot Light	O:4/2	PL1
Green - Run Pilot Light	O:4/3	PL2

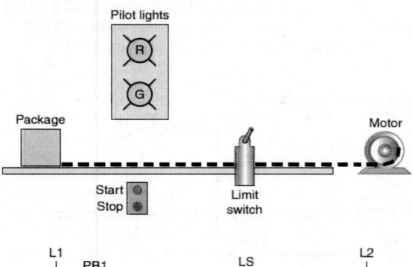

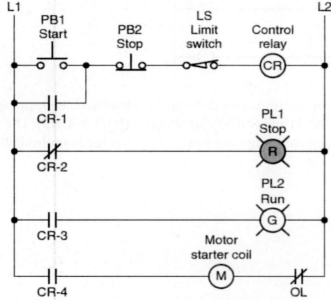

6-23 Write a program that will implement the drilling operation shown. The operation requires the drill press to turn "on" only if there is a part present and the operator has one hand on each of the start switches. This precaution is to ensure that the operator's hands are not in the way of the drill. Use the I/O Simulator screen and the following addresses to simulate the program:

PB1 (NO) _ I:3/4
PB2 (NO) _ I:3/5
Part Sensor (NO) _ I:3/6
Motor Contactor _ O:4/0

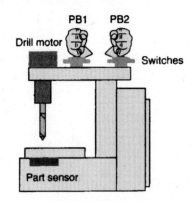

Sketch of the process

6-24 A temperature control system consists of four thermostats controlling three heating units. The thermostat contacts are set to close at 50°F, 60°F, 70°F, and 80°F, respectively. The PLC program is to be designed so that, at a temperature below 50°F, three heaters are to be "on." Between 50°F and 60°F, two heaters are to be "on." For 60°F to 70°F, one heater is to be "on." Above 80°F, there is a safety shutoff for all three heaters in case one stays on due to a malfunction. A master switch is to be used to turn the system on and off. Prepare a typical PLC program for this control process using the I/O Simulator screen and the following addresses to simulate the program:

50° Thermostat _ I:1/0
60° Thermostat _ I:1/1
70° Thermostat _ I:1/2
80° Thermostat _ I:1/3
Master Switch (NO) _ I:1/4
Heater #1 _ O:2/0
Heater #2 _ O:2/1
Heater #3 _ O:2/2

6-25 A pump is to be used to fill two storage tanks. The pump is manually started by the operator from a start/stop station. When the first tank is full, the control logic must be able to automatically stop flow to the first tank and direct flow to the second tank through the use of sensors and electric solenoid valves. When the second tank is full, the pump must shut down automatically. Indicator lamps are to be included to signal when each tank is full.

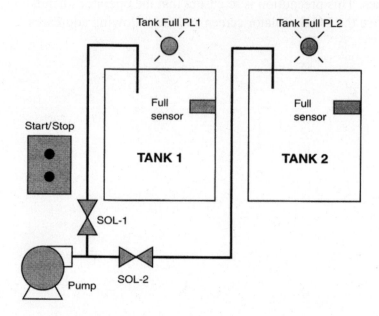

Prepare a typical PLC program for this control process using the I/O Simulator screen and the following addresses to simulate the program:

Stop PB (NC) _ I:1/0
Start PB (NO) _ I:1/1
Tank-1 Full Sensor (NO) _ I:1/2
Tank-2 Full Sensor (NO) _
I:1/3 _ Pump Motor _ O:2/0
SOL-1 (tank-1) _ O:2/1
SOL-2 (tank-2) _ O:2/2
Tank 1 Full PL1 _ O:2/3
Tank 2 Full PL2 _ O:2/4

6-26 Write the optimum ladder logic rung for the following scenario, and arrange the instructions for optimum performance. Turn "on" an output when switches SW6, SW7, and SW8, are all "true," or when SW55 is "true." Assume SW55 is an indication of an alarm state, so it is rarely "true"; SW7 is "true" most often, then SW8, then SW6. Use the I/O Simulator screen and the following addresses to simulate this rung:

SW6 (NO) _ I:1/0
SW7 (NO) _ I:1/1
SW8 (NO) _ I:1/2
SW55 (NO) _ I:1/3
Output _ O:2/0

6-27 Write a documented PLC program that will implement the reversing motor control schematic shown.

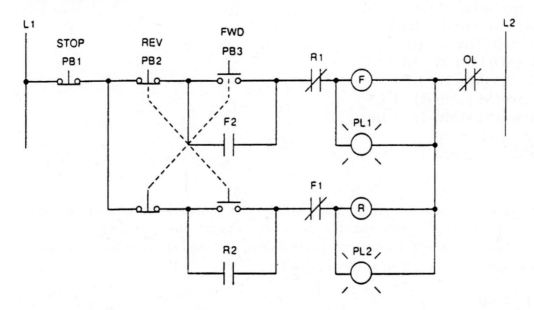

Use the I/O Simulator screen and the following addresses to simulate the program:

STOP (NC) _ I:1/0
FWD (NO) _ I:1/1
REV (NO) _ I:1/2
OL (NC) _ I:1/3
F _ O:2/0
R _ O:2/1
PL1 _ O:2/2
PL2 _ O:2/3

6-28 Write a documented PLC program that will implement the following motor control process:

➢ Two starters are to be wired so that each is operated from its own start/stop pushbutton station.
➢ A Master Stop switch is to be included that will trip out both starters when the switch is opened.
➢ Overload relay contracts are to be programmed so that an overload on any one of the starters will automatically drop out both of the starters.

Use the I/O Simulator screen and the following addresses to simulate the program:

Master Stop Switch (NO) _ I:1/0
Start PB (NO) (Mtr-1) _ I:1/1
Stop PB (NC) (Mtr-1) _ I:1/2
Start PB (NO) (Mtr-2) _ I:1/3
Stop PB (NC) (Mtr-2) _ I:1/4
OL Contact (NC) (Mtr-1) _ I:1/5
OL Contact (NC) (Mtr-2) _ I:1/6
Starter (Mtr-1) _ O:2/0
Starter (Mtr-2) _ O:2/1

6-29 Write a documented PLC program that will turn "on" a pilot light if one or the other of two switches is closed. If both switches are closed simultaneously, an alarm operates that can only be turned "off" by pushing a reset button. Use the I/O Simulator screen and the following addresses to simulate the program:

Switch-1 _ I:1/0
Switch-2 _ I:1/1
Reset PB (NC) _ I:1/2
PL _ O:2/0
Alarm _ O:2/1

6-30 Write a documented PLC program that will implement the hardwired control circuit shown. The control scheme is as follows:

➢ A person is required to actuate two pushbuttons, one at a time in the proper sequence, to operate a particular machine.
➢ The correct sequence is PB1 followed by PB2.
➢ Once the two buttons have been pressed in the correct sequence, the *correct sequence light* will come "on" and stay "on" until the reset button is pressed.
➢ If the wrong sequence occurs, the *error light* comes "on" and stays "on" until the reset button is pressed.
➢ Once an error has occurred, no sequence of operation of the pushbuttons can turn the machine "on."

Use the I/O Simulator screen and the following addresses to simulate the program:

PB1 _ I:1/0
PB2 _ I:1/1
Reset PB (NC) _ I:1/2
Correct Sequence PL1 _ O:2/0
Error PL2 _ O:2/1

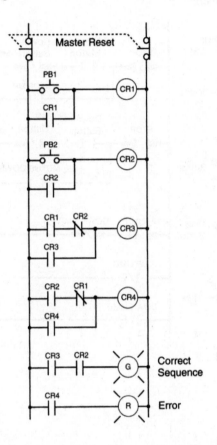

6-31 Implement the motorized door PLC program shown using the Door Simulator screen and the following addresses:

Up Limit _ I:1/3
Down Limit _ I:1/4
Up Button _ I:1/0
Down Button _ I:1/1
Stop _ I:1/2
Door Ajar _ O:2/2
Door Open _ O:2/3
Door Closed _ O:2/4
Motor Up _ O:2/0
Motor Down _ O:2/1

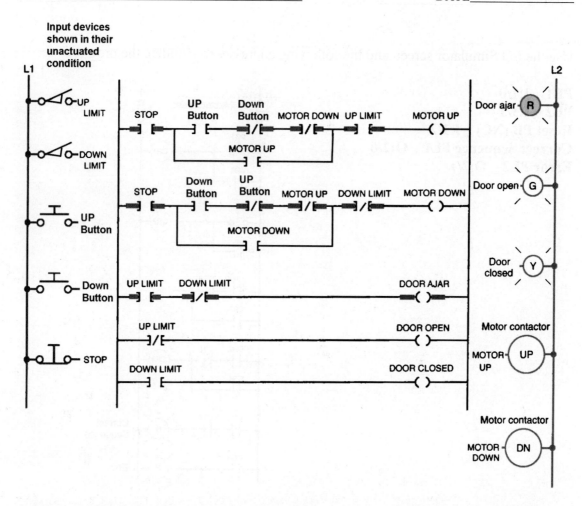

6-32 Write a documented PLC program that will implement the relay schematic shown. This circuit requires that Motor No. 2 cannot be started unless Motor No. 1 is running. Use the I/O Simulator screen and the following addresses to simulate the program:

Stop (1M) _ I:1/0
Start (1M) _ I:1/1
Stop (2M) _ I:1/2
Start (2M) _ I:1/3
1M _ O:2/0
PL1 _ O:2/1
2M _ O:2/2
PL2 _ O:2/3

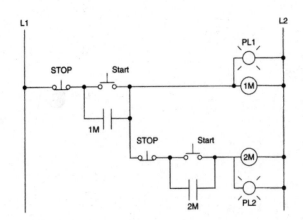

6-33 Write a documented PLC program that will implement the jogging control relay schematic shown. Use the I/O Simulator screen and the following addresses to simulate the program:

Stop (NO Button) _ I:1/0
Run (NO Button) _ I:1/1
Jog (NO Button) _ I:1/2
M _ O:2/0

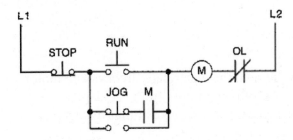

6-34 Write a documented PLC program that will execute the hardwired control circuit shown. Use the I/O Simulator screen and the following addresses to simulate the program:

PB1 (Start) _ I:1/0
PB2 (Stop) _ I:1/1
PS1 _ I:1/2
LS1 _ I:1/3
SS1 _ I:1/4
PL1 _ O:2/0
SOL 1 _ O:2/1
SOL 2 _ O:2/2
SOL 3 _ O:2/3
PL2 _ O:2/4

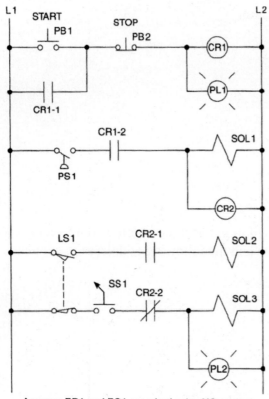

Assume: PB1 and PS1 are wired using NO contacts
LS1 is wired using one set of NC contacts

6-35 Write a documented program that will implement the electrically interlocked motor control circuit shown. This circuit is designed to avoid the motors from accidently operating in an order other than their proper sequence. Use the I/O Simulator screen and the following addresses to simulate the program:

Motor 1 Stop PB _ I:1/1   **M1 _ O:2/0**
Motor 2 Stop PB _ I:1/3   **M2 _ O:2/1**
Motor 3 Stop PB _ I:1/5   **M3 _ O:2/2**
Motor 1 Start PB - I:1/0
Motor 2 Start PB - I:1/2
Motor 3 Start PB - I:1/4

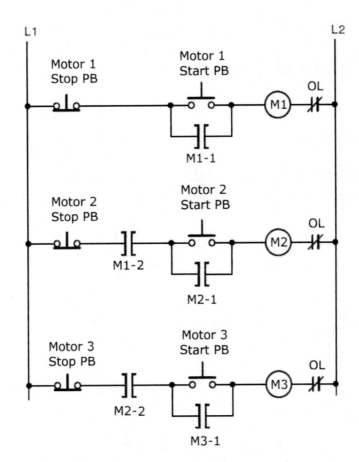

6-36 Write a documented program that will implement the hardwired pushbutton interlocking control circuit shown. This circuit is designed to prevent solenoid SOL-A and SOL-B from being energized at the same time. Use the I/O Simulator screen and the following addresses to simulate the program:

SOL-A PB- _ I:1/0   **SOL-A _ O:2/0**
SOL-B PB- _ I:1/1   **SOL-B _ O:2/1**

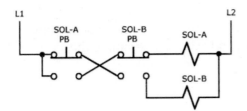

CHAPTER **7**

Programming Timers

LogixPro Programming Assignments

7-1 Write a documented program that will simulate the on-delay relay timer schematic shown. Use the I/O Simulator screen and the following addresses to simulate the program:

S1 _ I:1/0
L1 _ O:2/0
TON _ T4:0

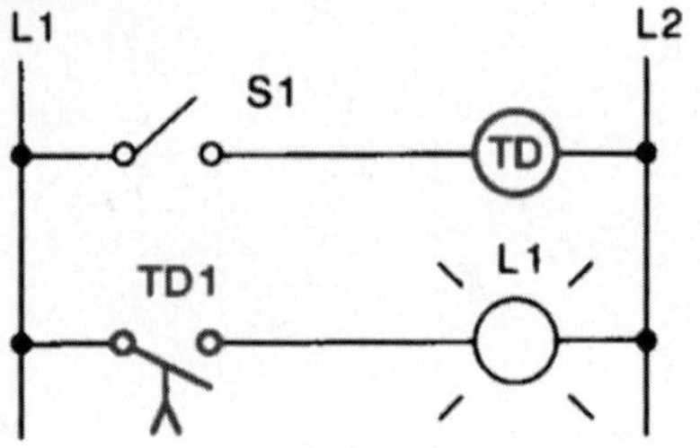

Sequence of operation:
S1 open, TD de-energized, TD1 open, L1 off.

S1 closes, TD energizes, timing period starts,
TD1 is still open, L1 is still off.

After 10 s, TD1 closes, L1 is switched on.

S1 is opened, TD de-energizes, TD1 opens instantly,
L1 is switched off.

7-2 Write a documented program that will simulate the off-delay relay timer schematic shown. Use the I/O Simulator screen and the following addresses to simulate the program:

S1 _ I:1/0
L1 _ O:2/0
TON _ T4:0

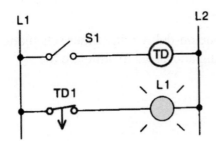

Sequence of operation:
S1 open, TD de-energized, TD1 closed, L1 on.

S1 closes, TD energizes, TD1 opens instantly,
L1 is switched off.

S1 is opened, TD de-energizes, timing period starts,
TD1 is still open, L1 is still off.

After 10 s, TD1 closes, L1 is switched on.

7-3 Write a documented program that will simulate the off-delay relay timer schematic shown. Use the I/O Simulator screen and the following addresses to simulate the program:

S1 _ I:1/0
L1 _ O:2/0
TON _ T4:0

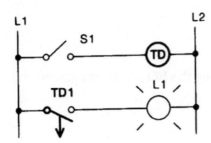

Sequence of operation:
S1 open, TD de-energized, TD1 open, L1 off.

S1 closes, TD energizes, TD1 closes instantly,
L1 is switched on.

S1 is opened, TD de-energizes, timing period starts,
TD1 is still closed, L1 is still on.

After 10 s, TD1 opens, L1 is switched off.

7-4 Write a documented program that will simulate the on-delay relay timer schematic shown. Use the I/O Simulator screen and the following addresses to simulate the program:

S1 _ I:1/0
L1 _ O:2/0
TON _ T4:0

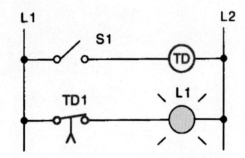

Sequence of operation:

S1 open, TD de-energized, TD1 closed, L1 on.

S1 closes, TD energizes, timing period starts,
TD1 is still closed, L1 is still on.

After 10 s, TD1 opens, L1 is switched off.

S1 is opened, TD de-energizes, TD1 closes instantly,
L1 is switched on.

7-5 Implement the on-delay PLC program shown using the I/O Simulator screen and the following addresses:

Input A _ I:1/0
Output B _ O:2/0
Output C _ O:2/1
Output D _ O:2/2
TON _ T4:0

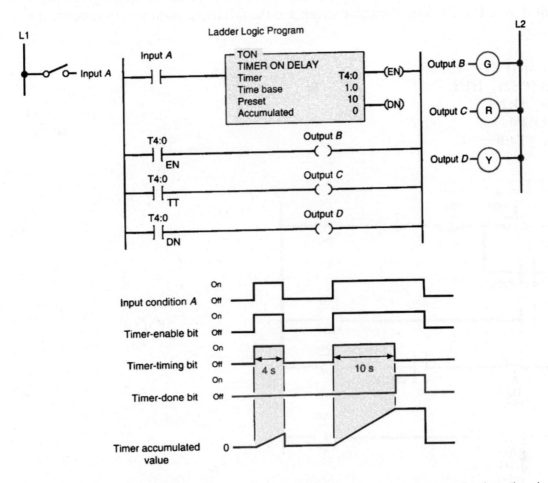

7-6 Write a documented program that will implement the motor control relay circuit shown. Use the Silo Simulator screen and the following addresses to simulate the program:

Start _ I:1/0
Stop _ I:1/1
M _ O:2/0
TON _ T4:0

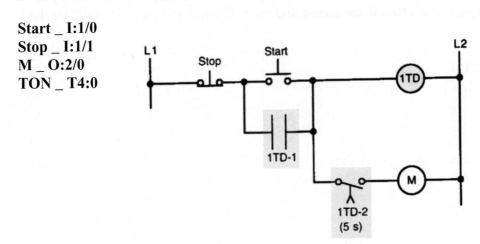

7-7 Write a documented program that will implement the start-up warning signal relay circuit shown. Use the Silo Simulator screen and the following addresses to simulate the program:

Start-up (PB1) _ I:1/0
Reset (PB2) _ I:1/1
Horn _ O:2/3
M _ O:2/0
TON _ T4:0

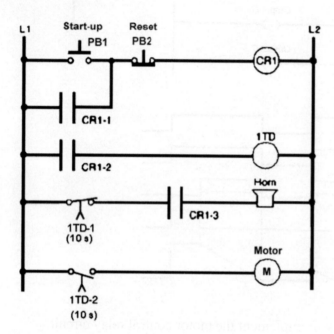

7-8 Write a documented program that will implement the automatic sequential relay circuit shown. Use the I/O Simulator screen and the following addresses to simulate the program:

Start (PB1) _ I:1/0
Stop (PB2) _ I:1/1
PS1 _ I:1/2
M1 _ O:2/0
M2 _ O:2/1
M3 _ O:2/2
TON _ T4:0

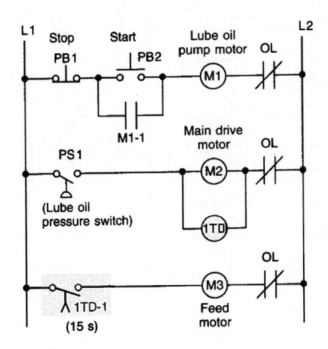

7-9 Implement the off-delay PLC program shown. In this application, closing the switch immediately turns "on" motors M1, M2, and M3. When the switch is opened, motors M1, M2, and M3 turn "off" at 5-second intervals. Use the I/O Simulator screen and the following addresses to simulate the program:

SW _ I:1/0
M1 _ O:2/0
M2 _ O:2/1
M3 _ O:2/2
TOF _ T4:1
TOF _ T4:2
TOF _ T4:3

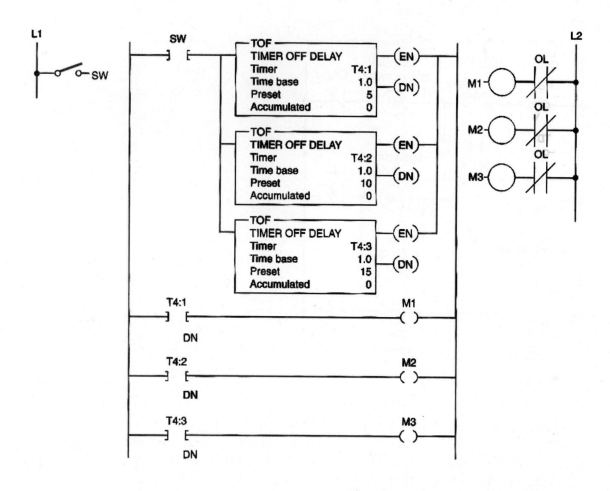

7-10 Write a documented program that will implement the pneumatic off-delay timer circuit shown. Use the I/O Simulator screen and the following addresses to simulate the program:

LS1 _ I:1/0
M1 _ O:2/0
M2 _ O:2/1
G _ O:2/2
R _ O:2/3
TOF _ T4:0

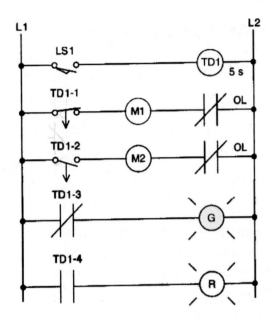

7-11 Implement the fluid pumping process PLC program shown. In this application:

> When the start button is pushed, the pump starts. The button can then be released, and the pump continues to operate.
> When the stop button is pushed, the pump stops.
> Before starting, PS1 must be closed.
> PS2 and PS3 must be closed 5 seconds after the pump starts. If either PS2 or PS1 opens, the pump will shut "off" and will not be able to start again for another 14 seconds.

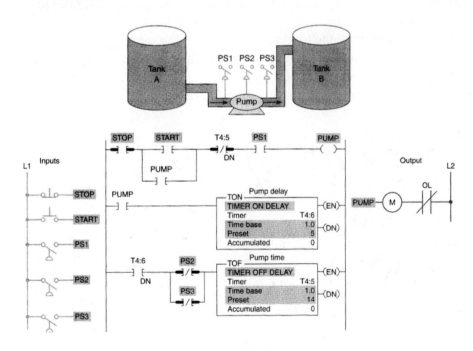

Use the I/O Simulator screen and the following addresses to simulate the program:

Stop _ I:1/0
Start _ I:1/1
PS1 _ I:1/2
PS2 _ I:1/3
PS3 _ I:1/4
M (pump) _ O:2/0
TOF _ T4:5
TON _ T4:6

7-12 Write a documented program for a retentive on-delay timer that will turn a light "on" anytime a switch is closed for an accumulated time of 15 seconds. Use an NO reset pushbutton to reset the timer. Use the I/O Simulator screen and the following addresses to simulate the program:

Switch _ I:1/0
Reset PB _ I:1/1
Light _ O:2/0
RTO _ T4:0

7-13 Implement the bearing lubrication PLC program shown. In this application:

➢ To start the machine, the operator turns SW "on."
➢ Before the motor shaft starts to turn, the bearings are supplied with oil by the pump for 10 seconds.
➢ When the operator turns SW "off" to stop the machine, the oil pump continues to supply oil for 15 seconds.
➢ A retentive timer is used to track the total running time of the pump. When the total running time is 3 hours, the motor is shut down and a pilot light is turned "on" to indicate that the filter and oil need to be changed.
➢ A reset button is provided to reset the process after the filter and oil have been changed.

Use the I/O Simulator screen and the following addresses to simulate the program:

Switch _ I:1/0
Reset PB _ I:1/1
Pump (M1) _ O:2/0
Motor (M2) _ O:2/1
PL _ O:2/2
TON _ T4:0
TOF _ T4:1
RTO _ T4:2

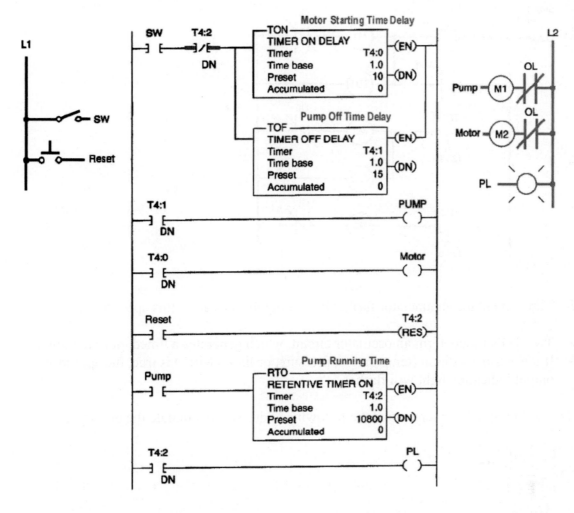

7-14 Write a documented program that will implement the sequential time delayed motor starting relay circuit shown. The three motors are started automatically in sequence with a 20-second time delay between each start-up. Use the Bottle Line or the I/O Simulator screen and the following addresses to simulate the program:

Stop PB1 _ I:1/0
Start PB2 _ I:1/1
M1 _ O:2/0
M2 _ O:2/1
M3 _ O:2/2
TON _ T4:0
TON _ T4:1

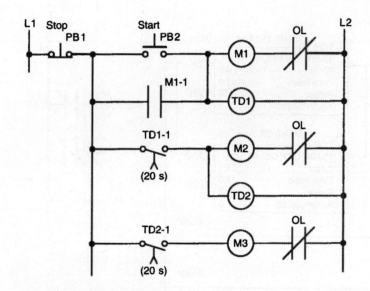

7-15 Implement the annunciator flasher PLC program shown. In this application:

➢ Two TON timers form an oscillator circuit, which generates a timed, pulsed output.
➢ If the alarm condition (temperature, pressure, or limit switch) is true, the appropriate output indicating light will flash.

Use the I/O Simulator screen and the following addresses to simulate the program:

TS1 _ I:1/0
PS1 _ I:1/1
LS1 _ I:1/1
G _ O:2/0
R _ O:2/1
Y _ O:2/2
TON _ T4:5
TON _ T4:6

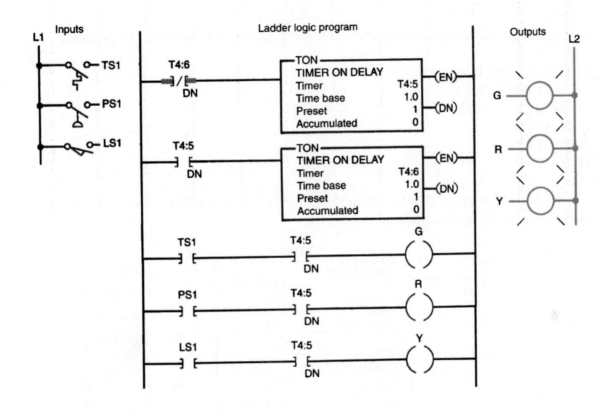

7-16 Write a documented program for two TON timers cascaded to give a longer time-delay period than the maximum preset time allowed for the single timer. Use the I/O Simulator screen and the following addresses to simulate the program:

SW _ I:1/0
PL _ O:2/0
TON _ T4:1
TON _ T4:2

7-17(a) Implement the control of traffic lights PLC program shown. In this application:

➢ The lights control the flow of traffic in one direction only.
➢ A cascading timer circuit accomplishes transition from red to green to amber.
➢ The sequence is Red 30 s "on," Green 25 s "on," and Amber 5 s "on."

Use the Traffic Simulator screen and the following addresses to simulate the program:

Red Light _ O:2/0
Amber Light _ O:2/1
Green Light _ O:2/2
TON _ T4:0
TON _ T4:1
TON _ T4:2

7-17(b) Add to the original program a pedestrian Walk light **(O:2/3)** that comes "on" anytime the traffic light is Red.

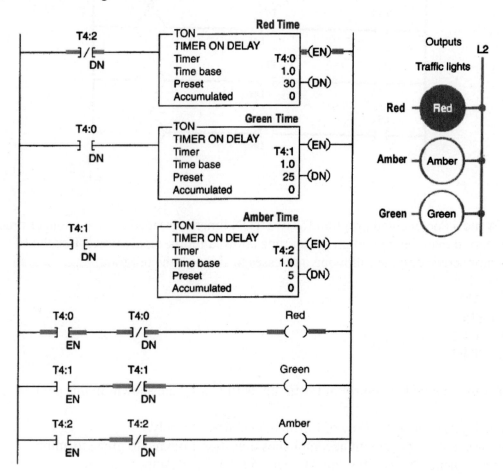

7-18(a) Implement the control of traffic lights in two directions PLC program shown.

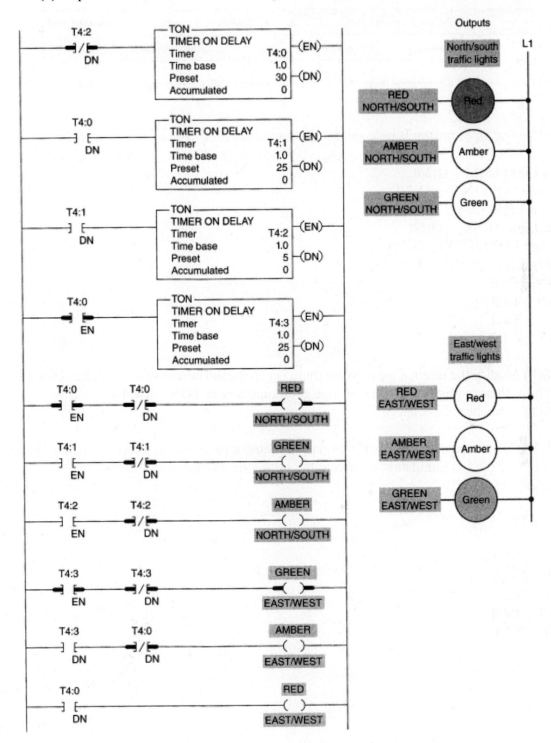

Timing chart

Red = north/south		Green = north/south	Amber = north/south
Green = east/west	Amber = east/west	Red = east/west	

◄——— 25 s ———►◄— 5 s —►◄——— 25 s ———►◄— 5 s —►

Use the Traffic Simulator screen and the following addresses to simulate the program:

Red Light (N/S) _ O:2/0
Amber Light (N/S) _ O:2/1
Green Light (N/S) _ O:2/2
Red Light (E/W) _ O:2/4
Amber Light (E/W) _ O:2/5
Green Light (E/W) _ O:2/6
TON _ T4:0
TON _ T4:1
TON _ T4:2
TON _ T4:3

7-18(b) Modify the original program so there is a 2-second period when both directions will have their red lights illuminated. Use additional timers **TON _ T4:4** and **TON _ T4:5** to implement this change.

7-19 Write a documented program that will implement the hardwired relay timer circuit shown. Use the I/O Simulator screen and the following addresses to simulate the program:

S1 _ I:1/0
PL1 _ O:2/0
PL2 _ O:2/1
PL3 _ O:2/2
PL4 _ O:2/3
TON _ T4:0
TOF _ T4:1

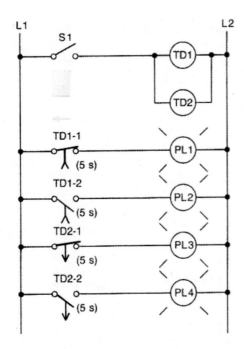

7-20 Write a documented program that will implement the hardwired relay control circuit shown. Use the I/O Simulator screen and the following addresses to simulate the program:

Hand/Auto _ I:1/0
Stop _ I:1/1
Start _ I:1/2
PS1 _ I:1/3
M1 _ O:2/0
TON _ T4:0

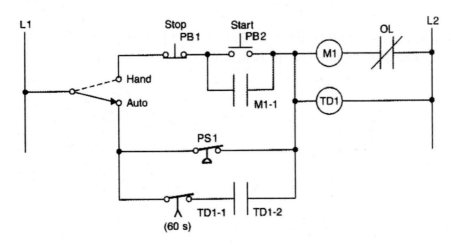

7-21 Implement the on-delay PLC program shown. Use the I/O Simulator screen and the following addresses to simulate the program:

LS1 _ I:1/0
SOL A _ O:2/0
SOL B _ O:2/1
R (PL) _ O:2/2
Y (PL) _ O:2/3
TON _ T4:0

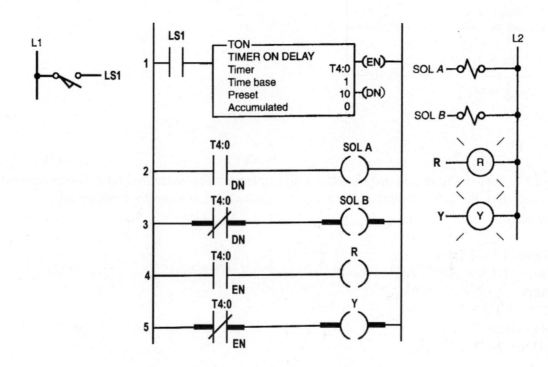

7-22 Implement the retentive on-delay PLC program shown. Use the I/O Simulator screen and the following addresses to simulate the program:

PB1 _ I:1/0
PB2 _ I:1/2
PL1 _ O:2/0
PL2 _ O:2/1
PL3 _ O:2/2
PL4 _ O:2/3
RTO _ T4:0

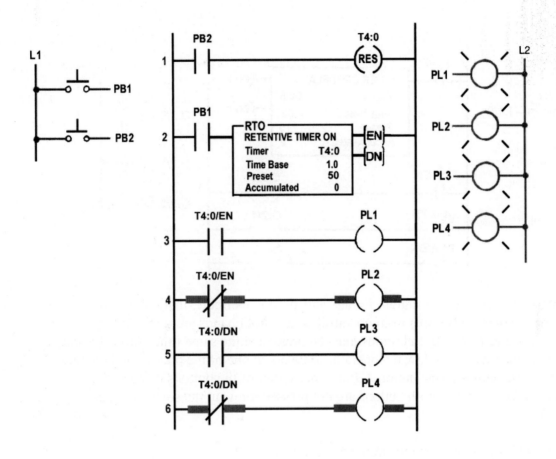

7-23 Write a program that will turn "on" a light PL **(O:2/0),** 15 seconds after switch S1 **(I:1/0)** has been turned "on." Use the I/O Simulator screen to simulate the program.

7-24 Implement the off-delay PLC program shown using the I/O Simulator screen and the following addresses:

S1 _ I:1/0
PL1 _ O:2/0
PL2 _ O:2/1
PL3 _ O:2/2
TOF _ T4:0

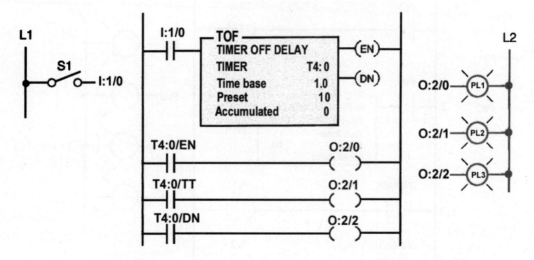

7-25 Write a documented program for an "anti-tie-down circuit" that will disallow a punch press solenoid **(O:2/0)** from operating unless both hands are on the two palm start buttons **(I:1/0 & I:1/1).** Both buttons must be pressed at the same time within 1 second. The circuit also will not allow the operator to tie down one of the buttons and operate only the press with just one button. (Hint: Once either of the buttons is pressed, begin timing 1 second. Then if both buttons are not pressed, prevent the solenoid from operating.)

Use the I/O Simulator screen to simulate the program.

7-26 Write a documented program to implement the process illustrated using the Batch Simulator. The sequence of operation is to be as follows:

➢ Normally open start and normally closed stop pushbuttons are used to start and stop the process at any time.
➢ When the start button is pressed, pump 1 energizes to start filling the tank.
➢ As the tank fills, the low-level sensor switch closes.
➢ When the tank is full, the high-level sensor switch closes and pump 1 is de-energized.
➢ The mixer motor then starts automatically and runs for a total of 3 minutes to mix the liquid.
➢ When the mixer motor stops, discharge pump 3 is energized to empty the tank.
➢ When the tank is empty, the low sensor switch opens to de-energize the discharge pump 3.
➢ The start button is pressed to repeat the sequence.

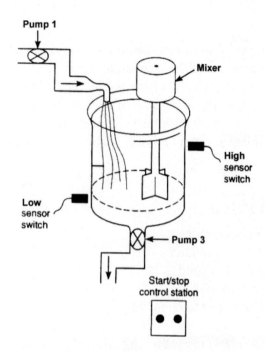

Pump 1

Mixer

High sensor switch

Low sensor switch

Pump 3

Start/stop control station

Use the Batch Simulator screen and the following addresses to simulate the program:

Start PB _ I:1/0
Stop PB _ I:1/1
Low-Level Sensor _ I:1/3
High-Level Sensor _ I:1/4
Mixer _ O:2/0
Pump1 _ O:2/1
Pump3 _ O:2/3
Internal _ B3:0/1
Timer-RTO _ T4:0

7-27 When the lights are turned "off" in a building, an exit door light is to remain "on" for an additional 2 minutes, and a parking lot light is to remain "on" for an additional 3 minutes after the door light goes out. Write a documented program to implement this process. Use the I/O Simulator screen and the following addresses to simulate the program:

Switch _ I:1/0
Building Lights _ O:2/0
Exit door Light _ O:2/1
Parking Lot Light _ O:2/2
TOF _ T4:0
TOF _ T4:1

7-28 Write a documented program to simulate the operation of a sequential taillight system. The light system consists of three separate lights on each side of the car. Each set of lights will be activated separately, by either the left or right turn signal switch. There is to be a 1-second delay between the activation of each light, and a 1-second period when all lights are "off." Ensure that with both turn signal switches "on," the system will not operate. Use the least number of timers possible. The sequence of operation should be as follows:

➢ The switch is operated.
➢ Light 1 is illuminated.
➢ Light 2 is illuminated 1 second later.
➢ Light 3 is illuminated 1 second later.
➢ Light 3 is illuminated for 1 second.
➢ All lights are "off" for 1 second.
➢ The system repeats while the switch is "on."

Use the I/O Simulator screen and the following addresses to simulate the program:

R-Switch _ I:1/0
L-Switch _ I:1/1
R-L1 _ O:2/0
R-L2 _ O:2/1
R-L3 _ O:2/2
L-L1 _ O:2/3
L-L2 _ O:2/4
L-L3 _ O:2/5
TON _ T4:1
TON _ T4:2
TON _ T4:3
TON _ T4:4

7-29 When a switch is turned "on," PL1 goes on immediately and PL12 goes on 9 seconds later. Opening the switch turns both lights "off." Write a documented program that will implement this process. Use the I/O Simulator screen and the following addresses to simulate the program:

Switch _ I:1/0
PL1 _ O:2/0
PL12 _ O:2/1
TON _ T4:0

7-30 When a switch is turned on, PL1 and PL2 immediately come on. When the switch is turned "off," PL1 immediately goes "off." PL2 remains "on" for another 3 seconds and then goes "off." Write a documented program that will implement this process. Use the I/O Simulator screen and the following addresses to simulate the program:

Switch _ I:1/0
PL1 _ O:2/0
PL2 _ O:2/1
TOF _ T4:0

7-31 Write a documented program that will turn "on" pilot light PL1 10 seconds after switch S1 is turned "on." Pilot light PL2 will come "on" 5 seconds after PL1 comes "on." Pilot light PL3 will come "on" 8 seconds after PL2 comes "on." Pressing normally closed pushbutton PB1 will reset all timers but only if PL3 is "on." Use the I/O Simulator screen and the following addresses to simulate the program:

Switch _ I:1/0
PB1 (NC) _ I:1/1
PL1 _ O:2/0
PL2 _ O:2/1
PL3 _ O:2/2
TON _ T4:1
TON _ T4:2
TON _ T4:3

7-32 When a switch is turned "on," PL1 and PL2 immediately come "on." PL1 turns "off" after 4 seconds. PL2 remains "on" until the switch is turned "off." Turning the switch "off" at any time turns both lights "off." Write a documented program that will implement this process. Use the I/O Simulator screen and the following addresses to simulate the program:

Switch _ I:1/0
PL1 _ O:2/0
PL2 _ O:2/1
TON _ T4:0

7-33 Write a documented program for a display sign that will sequentially turn "on" three lights 2 seconds apart, then turn all three lights "off," and repeat the sequence. Use the I/O Simulator screen and the following addresses to simulate the program:

ON/OFF Switch _ I:1/0
L1 _ O:2/0
L2 _ O:2/1
L3 _ O:2/2
TON _ T4:1
TON _ T4:2
TON _ T4:3
TON _ T4:4

7-34 A saw, fan, and lube pump all turn "on" when a start button is pressed. Pressing a stop button immediately stops the saw but allows the fan to continue operating. The fan is to run for an additional 5 seconds after shutdown of the saw. If the saw has operated for more than 20 seconds, the fan should remain "on" until reset by a separate fan reset button. If the saw has operated less than 20 seconds, the lube pump should go "off" when the saw is turned "off." However, if the saw has operated for more than 20 seconds, the lube pump should remain "on" for an additional 10 seconds after the saw is turned "off." Write a documented program that will implement this process. Use the I/O Simulator screen and the following addresses to simulate the program:

Start PB1 (NO) _ I:1/3
Stop PB2 (NC) _ I:1/4
Reset PB (NC) _ I:1/5
Saw _ O:2/0
Fan _ O:2/1
Lube Pump _ O:2/2
TON _ T4:1
TON _ T4:2
TON _ T4:3

7-35 Write a documented program that will execute the relay timer schematic shown. Use the I/O Simulator screen and the following addresses to simulate the program:

Stop PB1 (NC) _ I:1/0
Start PB2 (NO) _ I:1/1
L1 _ O:2/0
Internal _ B3:0/1
TON _ T4:1
TON _ T4:2

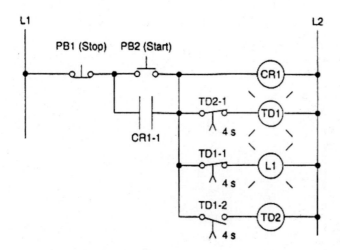

7-36 Write a documented program that will implement the relay schematic shown. The control process consists of three motors: M1, M2, and M3. The electrical control system is to be designed so that motor M1 must be running before motor M2 or M3 can be started. Each motor has its own start/stop pushbutton station. Both motors M2 and M3 can normally be stopped or started without affecting the operation of motor M1. However, if all three motors are running, the stopping of any one motor, for any reason, will automatically stop all three motors. Use the I/O Simulator screen and the following addresses to simulate the program:

Stop (M1) _ I:1/0
Start (M1) _ I:1/1
Stop (M2) _ I:1/2
Start (M2) _ I:1/3
Stop (M3) _ I:1/4
Start (M3) _ I:1/5
M1 _ O:2/0
M2 _ O:2/1
M3 _ O:2/2
Internal _ B3:0/1
TOF _ T4:0

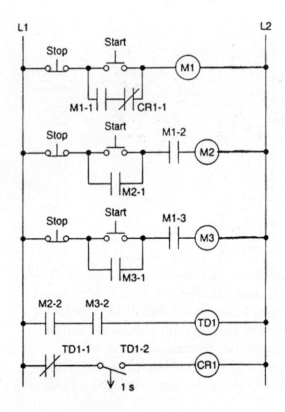

7-37 Traffic flow on a one-way street is to be controlled by means of a pedestrian pushbutton so that the GREEN traffic light and DON'T walk indicator are to be normally "on" at all times when the pedestrian pushbutton is not actuated. When the pedestrian pushbutton is actuated, the timing is started and controls the outputs as follows:

➢ The GREEN traffic light immediately switches "off" and the AMBER traffic light switches "on" to begin to stop the traffic flow. The outputs remain in this state for 5 seconds.
➢ The AMBER traffic light switches "off" and the RED traffic light switches "on." The outputs remain in this state for 5 seconds, to ensure that traffic has stopped before pedestrians begin to cross.
➢ The WALK pedestrian light switches "on" and the RED traffic light remains "on." Outputs remain in this state for 15 seconds, allowing pedestrians safe passage across the street.
➢ The WALK pedestrian light switches "off" and the RED traffic light remains "on." Outputs remain in this state for 5 seconds, to ensure that pedestrians are not still crossing the street when the traffic light changes from RED to GREEN.
➢ The GREEN traffic light switches "on" and the RED traffic light switches "off." Outputs remain in this state for at least 30 seconds to ensure a minimum amount of automotive traffic flow time even if the walk pushbutton is pressed to start the cycle again.

Use the Traffic Simulator screen and the following addresses to simulate the program:

Pedestrian Pushbutton (NO) _ I:1/0
RED Traffic Light _ O:2/0
AMBER Traffic Light _ O:2/1
GREEN Traffic Light _ O:2/2
Pedestrian WALK Light _ O:2/3
Internal _ B3:1/0
Internal _ B3:1/1
TON _ T4:1
TON _ T4:2
TON _ T4:3
TON _ T4:4
TON _ T4:5

7-38 Modify the continuous filling Silo simulation program shown. The modification calls for:

➤ A 2-second time-delay period preceding the filling of the box after the conveyor has stopped.
➤ A 4-second time-delay period preceding the starting of the conveyor after the box is filled.

Use the Silo Simulator screen and the following addresses to simulate the program:

Start (NO contact) _ I:1/0
Stop (NC contact) _ I:1/1
Photo Switch (NO contact) _ I:1/3
Level Switch (NO contact) _ I:1/4
Run Light _ O:2/2
Standby Light _ O:2/3
Full Light _ O:2/4
Internal _ B3:1/0
TON _ T4:1
TON _ T4:2
Conveyor Motor _ O:2/0
Solenoid _ O:2/1

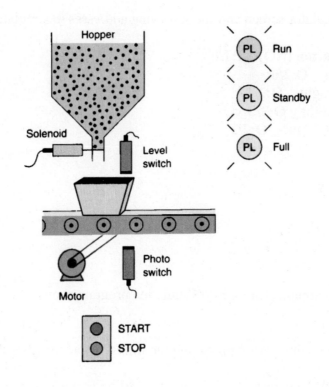

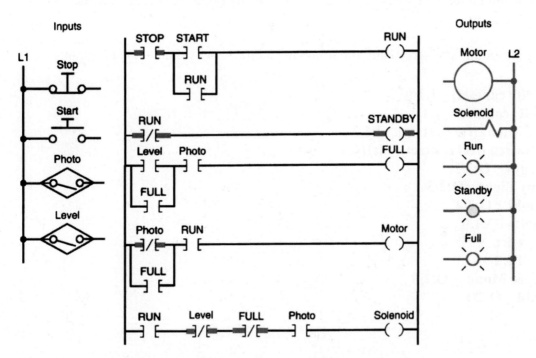

CHAPTER **8**

Programming Counters

LogixPro Programming Assignments

8-1 Implement the up-counter PLC program shown using the I/O Simulator screen and the addresses given.

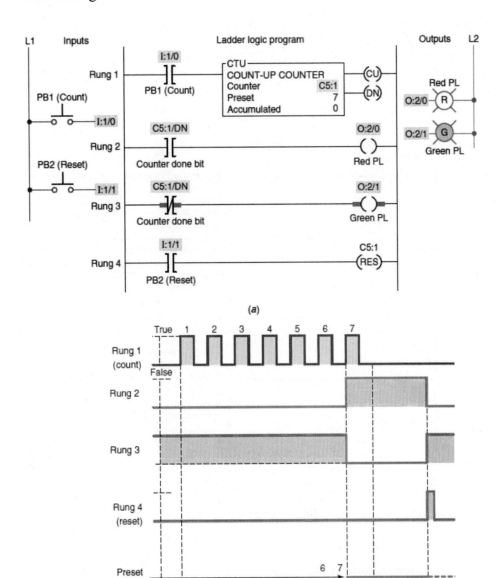

(a)

8-2 Write a documented program that will simulate the parts counting program shown. Counter C5:2 counts the total number of parts coming off of an assembly line for final packaging. Each package must contain 10 parts. When 10 parts are detected, counter C5:1 sets bit B3:1 to initiate the box closing sequence. Counter C5:3 counts the total number of packages filled in a day. (The maximum number of packages per day is 300.) A pushbutton is used to restart all the counters from zero daily. Use the Silo Simulator screen and the following addresses to simulate the program:

Conveyor Stop PB (NC) _ I:1/1
Conveyor Start PB (NO) _ I:1/0
Proximity Switch _ I:1/3
Reset Switch (B) _ I:1/6
Box Closing Sequence Bit _ O:2/4 (B3:1)
Conveyor Motor _ O:2/0
CTU _ C5:1
CTU _ C5:2
CTU _ C5:3

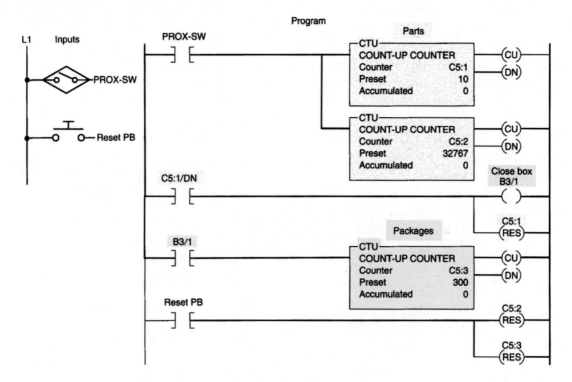

8-3 Write a documented program that will implement the following conveyor motor control process:

➢ The start button is pressed to start the conveyor motor.
➢ Cases move past the proximity switch and increment the counter's accumulated value.

➢ After a count of 13, the conveyor motor stops automatically and the counter's accumulated value is reset to zero.
➢ The conveyor motor can be stopped and started manually at any time without loss of the accumulated count.
➢ The accumulated count of the counter can be reset manually at any time by means of the count reset button.
➢ The process is repeated when the start button is pressed.

Use the Silo Simulator screen and the following addresses to simulate the program:

Conveyor Stop PB (NC) _ I:1/1
Conveyor Start PB (NO) _ I:1/0
Proximity Switch _ I:1/3
Count Reset Switch (B) _ I:1/6
Conveyor Motor _ O:2/0
Internal _ B3:0/0
CTU _ C5:0

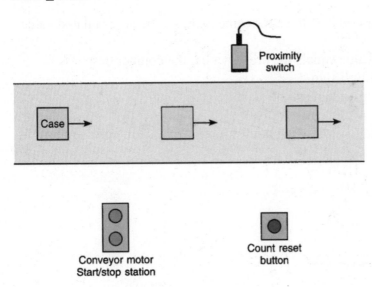

8-4 Write a documented program that will implement the following alarm monitor circuit:

➢ The closing of a high liquid level sensor switch triggers an alarm light.
➢ The light will flash whenever the alarm condition is triggered and has not been acknowledged, even if the alarm condition clears in the meantime.
➢ The alarm is acknowledged by closing a selector switch.
➢ The light will operate in the steady on mode when the alarm trigger condition still exists but has been acknowledged.

Use the Batch Simulator screen and the following addresses to simulate the program:

Fill (Start PB-NO) _ I:1/0
Empty (Stop PB-NC) _ I:1/1
Pump1 (Fill Pump) O:2/1
Pump3 (Discharge Pump) O:2/3
High-Level Sensor (NO) _ I:1/4
Selector Switch (NO Pos B) _ I:1/10
Alarm Light _ O:2/7
TON _ T4:5
TON _ T4:6
CTU _ C5:1

8-5(a) Implement the up/down-counter PLC program shown. This program is used to keep count of the cars that enter and leave a parking garage and operates as follows:

➢ As a car enters, it triggers the up-counter output instruction and increments the accumulated count by 1.
➢ As a car leaves, it triggers the down-counter output instruction and decrements the accumulated count by 1.
➢ Since both the up- and down-counters have the same address, the accumulated value will be the same in both.
➢ Whenever the accumulated value equals the preset value, the counter output is energized to light up the Lot Full sign.
➢ A reset button has been provided to reset the accumulated count.

Use the I/O Simulator screen and the following addresses to simulate the program:

Enter Proximity Switch (NO) _ I:1/0
Exit Proximity Switch (NO) _ I:1/1
Reset PB (NO) _ I:1/2
Lot Full Light _ O:2/0
Up/Down-Counter _ C5:1

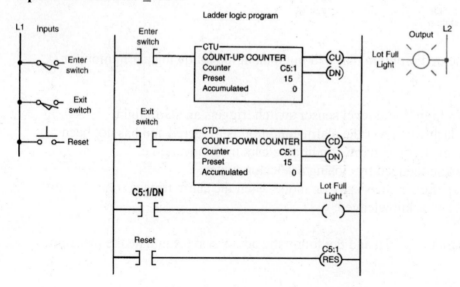

8-5(b) Add an additional up-counter **(C5:2)** and NO reset pushbutton **(I:1/3)** to the original counter to keep a running tally on the total number of cars that enter the parking garage for any given period.

8-6 Implement the up/down-counter PLC program shown. Use the I/O Simulator screen and the following addresses to simulate the program:

Input A _ **I:1/0**
Input B _ **I:1/1**
Input C _ **I:1/2**
Output A _ **O:2/0**
Output B _ **O:2/1**
Output C _ **O:2/0**
Up/Down-Counter _ **C5:2**

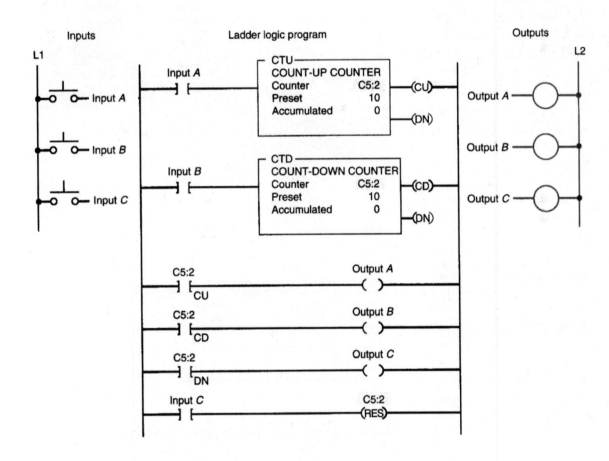

8-7(a) Implement the up/down-counter program used in an in-process monitoring system. This program is designed to provide continuous monitoring of items in-process and operates as follows:

➢ An in-feed photoelectric sensor counts raw parts going into the system.
➢ An out-feed photoelectric sensor counts finished parts leaving the machine.
➢ The number of parts between the in-feed and out-feed is indicated by the accumulated count of the counter.
➢ Before start-up, the system is completely empty of parts and the counter is reset manually to zero.
➢ The counter preset value is irrelevant in this application since the output on or off logic is not used.

Use the I/O Simulator screen and the following addresses to simulate the program:

In-Feed Count _ I:1/0
Out-Feed Count _ I:1/1
Reset Button _ I:1/2
Up/Down-Counter _ C5:1

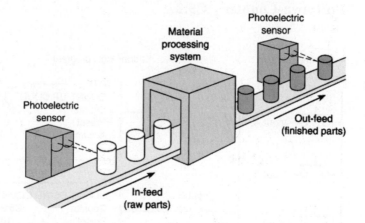

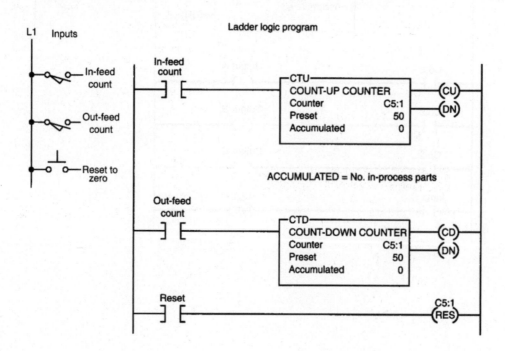

8-7(b) Add a pilot light **(PL1 _ O:2/0)** to the original program that will come on any time the number of parts in-process is 10 or greater. Use the latch and unlatch instruction to implement this added feature.

8-8 Write a documented program that demonstrates how to cascade two counters to count events that exceed the maximum number allowed per counter. Use the I/O Simulator screen and the following addresses to simulate the program:

Count Button (NO) PB1 _ I:1/0
Reset Button (NC) _ I:1/1
Up-Counter _ C5:0
Up-Counter _ C5:1

8-9 Implement the cascading of two counters, to store an extremely large number of counts, program shown. Use the I/O Simulator screen and the following addresses to simulate the program:

Count Button PB1 _ I:1/0
Reset Button PB2 _ I:1/1
Light _ O:2/0
Up-Counter _ C5:1
Up-Counter _ C5:2

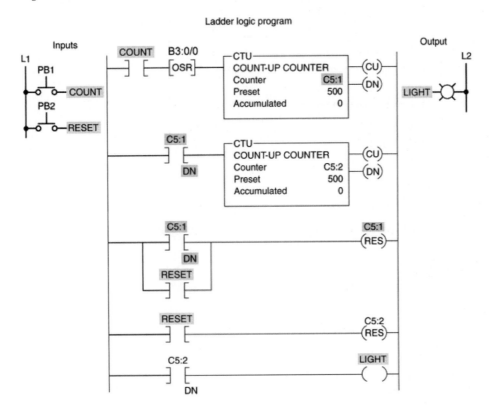

Ladder logic program

8-10 Implement the simple timer/counter program shown to simulate a time-of-day clock measuring time in hours, minutes, and seconds. Add an ON/OFF switch, and use the I/O Simulator screen and the following addresses to simulate the program:

ON/OFF Switch _ I:1/0
RTO _ T4:0
Up-Counter _ C5:0
Up-Counter _ C5:1

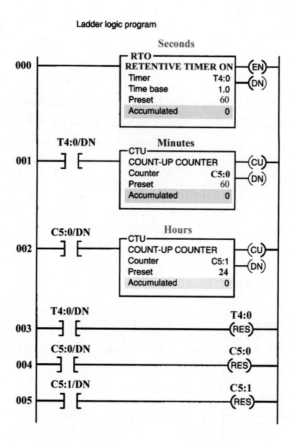

8-11 Implement the program shown, for monitoring the time of an event. Use the I/O Simulator screen and the following addresses to simulate the program:

Pressure Switch _ I:1/0
Reset _ I:1/1
Set Light _ O:2/0
Trip Light _ O:2/1
RTO _ T4:0
Internal Relay _ B3:0/0
Up-Counter _ C5:0
Up-Counter _ C5:1

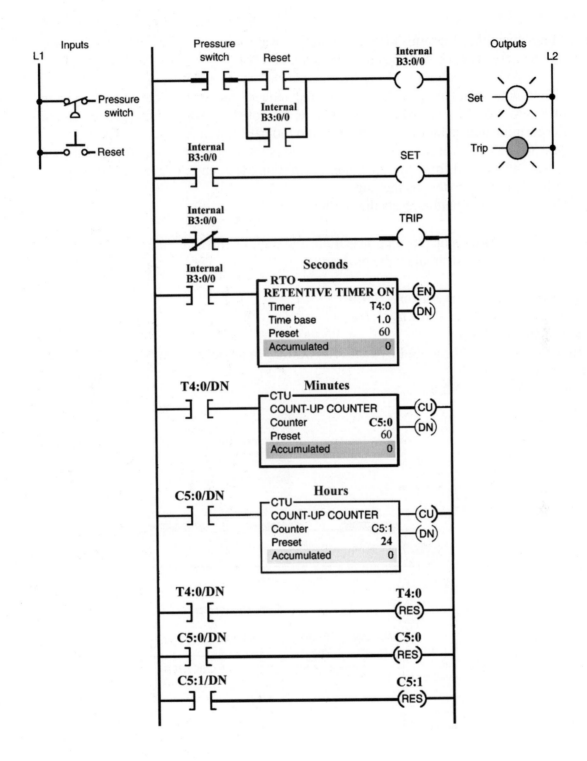

8-12 Implement the program shown of a counter used for length measurement. The operation of the program can be summarized as follows:

➤ Count input pulses are generated by the magnetic sensor, which detects passing teeth on a conveyor drive sprocket. If 10 teeth per foot of conveyor motion pass the sensor, the accumulated count of the counter would indicate feet in tenths.
➤ The photoelectric sensor monitors a reference point on the conveyor. When activated, it prevents the unit from counting, thus permitting the counter to accumulate counts only when the bar stock is moving.
➤ Closing the reset button resets the counter.

Add a conveyor start/stop pushbutton station, and use the Silo Simulator screen and the following addresses to simulate the program:

Conveyor Motor Start Button _ I:1/0
Conveyor Motor Stop Button _ I:1/1
Photo Sensor _ I:1/3
Magnetic Sensor _ Simulated by the free-running timer located in S:4/1
Reset _ I:1/7 (Pos C)
Conveyor Motor _ O:2/0
CTU _ C5:1

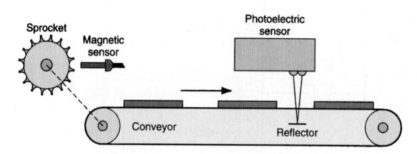

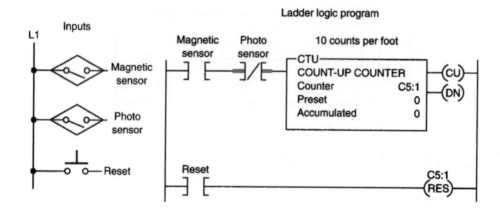

8-13 Implement the automatic stacking process shown. In this process, conveyor M1 is used to stack metal plates onto conveyor M2. The photoelectric sensor provides an input pulse to the PLC counter each time a metal plate drops from conveyor M1 to M2. When 15 plates have been stacked, conveyor M2 is activated for 5 seconds by the PLC timer. The operation can be summarized as follows:

➢ When the start button is pressed, conveyor M1 begins running.
➢ After 15 plates have been stacked, conveyor M1 stops and conveyor M2 begins running.
➢ After conveyor M2 has been operated for 5 seconds, it stops and the sequence is repeated automatically.
➢ The done bit of the timer resets the timer and counter, and provides a momentary pulse to automatically restart conveyor M1.

Use the bottles, instead plates, in conjunction with the Bottle Line Simulator screen and the following addresses to simulate the program:

Stop PB _ I:1/0
Start PB _ I:1/1
Conveyor M1 _ O:2/0
Conveyor M2 _ O:2/2
TON _ T4:1
CTU _ C5:1

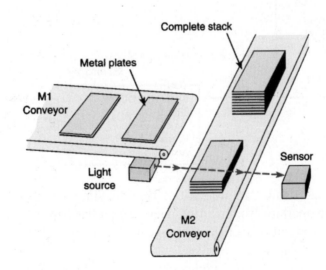

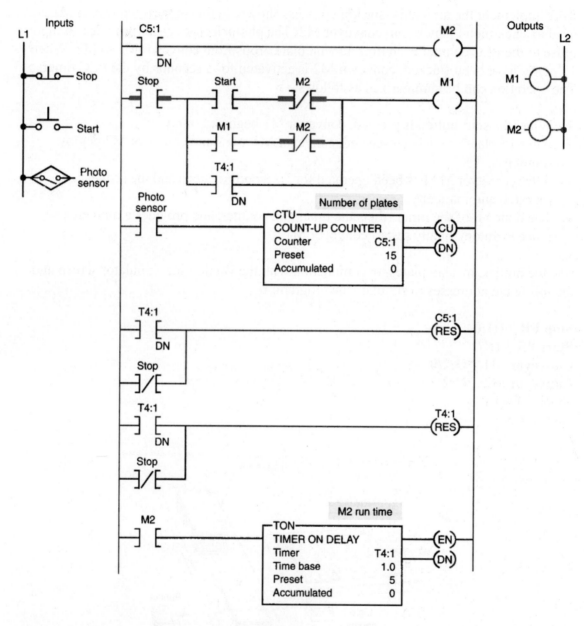

8-14 Implement the automatic motor lock-out program shown. This program is designed to prevent a machine operator from starting a motor that has tripped off more than five times in an hour. The operation of the program can be summarized as follows:

➤ The normally open overload (OL) relay contact momentarily closes each time an overload is sensed.
➤ Every time the motor stops due to an overload condition, the motor start circuit is locked out for 5 minutes.

> If the motor trips off more than five times in 1 hour, the motor start circuit is permanently locked out and cannot be started until the reset button is actuated.
> The lock-out pilot light is switched on whenever a permanent lock-out condition exists.

Use the Dual Compressor Simulator screen and the following addresses to simulate the program:

Stop PB _ I:1/0
Start PB _ I:1/1
OL Relay (NO) _ I:1/5 (Selector Switch Pos. B)
Reset (NO) _ I:1/6 (Selector Switch Pos. C)
Motor _ O:2/0
Lockout Light _ O:2/3
TON _ T4:0
TON _ T4:1
CTU _ C5:0

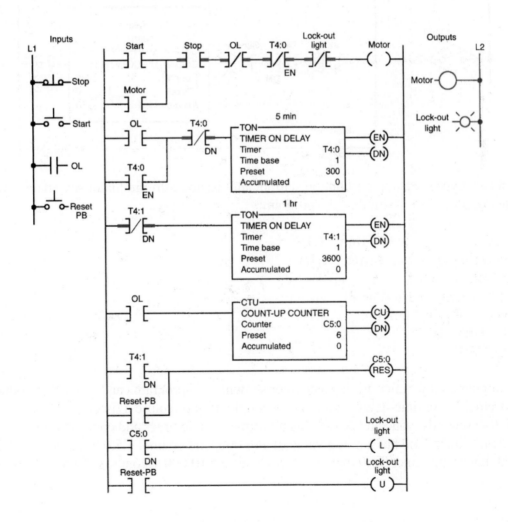

8-15 Implement the product flow rate program shown. This program is designed to indicate how many parts per minute pass a given process point. The operation of the program can be summarized as follows:

➢ When the start switch is closed, both the timer and counter are enabled.
➢ The counter is pulsed for each part passing the parts sensor.
➢ The counting begins and the timer starts timing through its 1-minute time interval.
➢ At the end of 1 minute, the timer done bit causes the counter rung to go false. Sensor pulses continue but do not affect the PLC counter. The number of parts per minute is represented by the accumulated value of the counter.
➢ The sequence is reset by momentarily opening and closing the start switch.

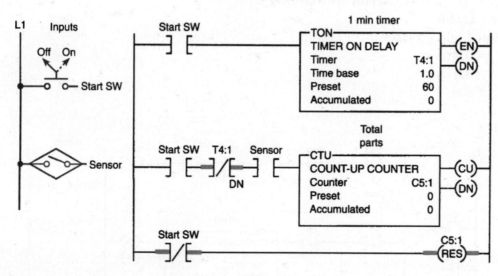

Ladder logic program

Add a conveyor start/stop pushbutton station, and use the Silo Simulator screen and the following addresses to simulate the program:

Conveyor Motor Start Button _ I:1/0
Conveyor Motor Stop Button _ I:1/1
Parts Sensor _ I:1/3
ON/OFF Switch _ I:1/6 (Pos B)
Conveyor Motor _ O:2/0
TON _ T4:1
CTU _ C5:1

8-16 Implement the timer driving a counter shown. This program can be used to produce an extremely long time-delay period. For example, if you require a timer to time to 1,000,000 seconds, you can achieve this by using a single timer and counter programmed as shown. Timer T4:0 has a preset value of 100,000, and counter C5:0 has a preset value of 100. Each time timer T4:0 input contact closes for 10,000 seconds, its done bit resets timer T4:0 and increments counter C5:0 by 1. When the done bit of timer T4:0 has turned on and off 100 times, the output light becomes energized. Therefore, the output light turns on after 10,000 × 100, or 1,000,000, seconds after the timer input contact closes. Use the I/O Simulator screen and the following addresses to simulate the program:

S1 Timer Input _ I:1/0
Light _ O:2/0
TON _ T4:0
CTU _ C5:0

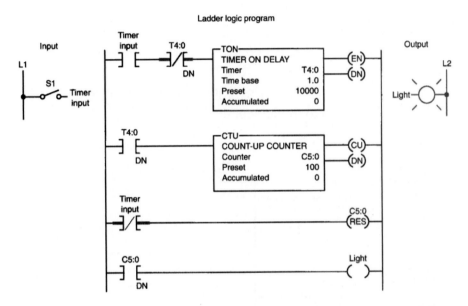

8-17 Program the counter program shown (Chapter 8, problem 1 of the text). Use the I/O Simulator screen and the addresses given to implement the program.

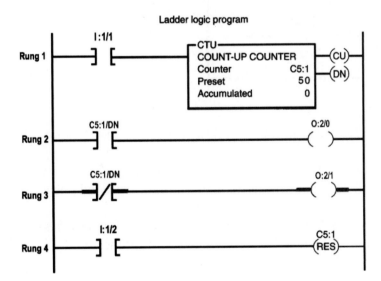

8-18 Program the PLC program shown (Chapter 8, problem 2 of the text). Use the I/O Simulator screen and input address **I:1/0** to implement the program.

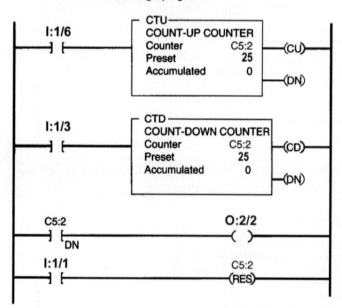

8-19 Program the counter program shown (Chapter 8, problem 3 of the text). Use the I/O Simulator screen and the addresses given to implement the program.

8-20 Write a documented PLC program for the following counter specification:

➤ Counts the number of times a pushbutton (PB1) is closed.
➤ Decrements the accumulated value of the counter each time a second pushbutton (PB2) is closed.
➤ Turns on a light (PL1) any time the accumulated value of the counter is less than 20.
➤ Turns on a second light (PL2) when the accumulated value of the counter is equal to or greater than 20.
➤ Resets the counter to zero when a selector switch (SS) is closed.

Use the I/O Simulator screen and the following addresses to simulate the program:

PB1 _ I:1/0
PB2 _ I:1/1
SS _ I:1/2
PL1 _ O:2/0
PL2 _ O:2/1
Counter _ C5:1

8-21 Write a documented PLC program that will execute the following control circuit correctly:

> ➤ Turns on a nonretentive timer when a switch (SW) is closed (preset value of timer is 10 seconds).
> ➤ Timer is automatically reset by a programmed transitional contact when it times out.
> ➤ Counter counts the number of times the timer goes to 10 seconds.
> ➤ Counter is reset automatically by a second programmed transitional contact at a count of 5.
> ➤ Latches on a light (PL1) at a count of 5.
> ➤ Resets light to off and counter to zero when a selector switch (SS) is closed.

Use the I/O Simulator screen and the following addresses to simulate the program:

SW _ I:1/1
SS _ I:1/2
PL1 _ O:2/0
Internal Relay _ B3:0
Timer _ T4:1
Counter _ C5:1

8-22 Write a documented PLC program that will execute the following industrial control process. The sequence of operation is as follows:

> ➤ Product in position (limit switch LS1 contacts close).
> ➤ The start button is pressed, and the conveyor motor starts to move the product forward toward position A (limit switch LS1 contacts open when the actuating arm returns to its normal position).
> ➤ The conveyor moves the product forward to position A and stops (position A is detected by eight off-to-on pulses from the encoder, which are counted by an up-counter).
> ➤ A time delay of 10 seconds occurs, after which the conveyor starts to move the product to LS2 and stops (LS2 contacts close when the actuating arm is hit by the product).

➢ An emergency stop button is used to stop the process at any time.
➢ If the sequence is interrupted by an emergency stop, counter and timer are reset automatically.

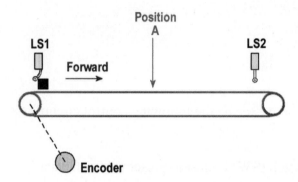

Use the I/O Simulator screen and the following addresses to simulate the program:

LS1 _ I:1/1
LS2 _ I:1/2
Start PB (NO) _ I:1/3
Encoder (NO PB) _ I:1.4
Emergency Stop (NO PB) _ I:1.5
Conveyor Motor _ O:2/0
Internal Relay _ B3:0
RTO Timer _ T4:2
Up-Counter _ C5:1

8-23 Write a documented PLC program to implement the process shown. An up-counter must be programmed as part of a batch-counting operation to sort parts automatically for quality control. The counter is installed to divert 1 part out of every 10 for quality control or inspection purposes. The circuit operates as follows:

➢ A start/stop pushbutton station is used to turn the conveyor motors (M1 and M2) on and off.
➢ A proximity sensor counts the parts as they pass by on the conveyor.
➢ When 10 is reached, the counter's output activates the gate solenoid, diverting the part to the inspection line.
➢ The gate solenoid is energized for 0.3 second, which allows enough time for the part to continue to the quality control line.
➢ The gate returns to its normal position when the 0.3-second time period ends.
➢ The counter resets to zero and continues to accumulate counts.
➢ A reset switch is provided to reset the counter manually.

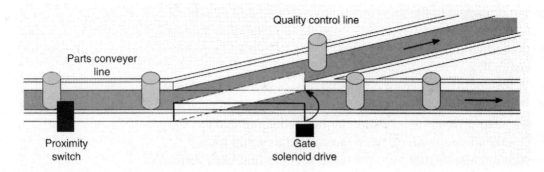

Use the Bottle Line Simulator screen and the following addresses to simulate the program:

Stop PB _ I:1/0
Start PB _ I:1/1
Proximity Sensor (NO) _ I:1/6
Reset (NO) _ I:1/3 (Selector Switch Pos. B)
Motor M1 _ O:2/0
Motor M2 _ O:2/2
Gate Solenoid _ O:2/5
TON _ T4:1
CTU _ C5:1

8-24 Write a documented PLC program that will increment a counter's accumulated value 1 count every 1 second. A second counter's accumulated value will increment 1 count every time the first counter's accumulated value reaches 10. The first counter will reset when its accumulated value reaches 10, and the second counter will reset when its accumulated value reaches 5. Use the I/O Simulator screen and the following addresses to simulate the program:

ON/OFF Switch _ I:1/0
TON _ T4:10
CTU (First Counter) _ C5:7
CTU (Second Counter) _ C5:8

8-25 Write a documented PLC program to implement the process shown. For this application, a company that makes electronic assembly kits needs a counter to count the number of resistors placed into each kit. The controller must stop the take-up spool at a predetermined amount of resistors (20). A worker on the floor will then cut the resistor strip and place it in the kit. The circuit operates as follows:

➢ A start/stop pushbutton station is used to turn the spool motor drive on and off manually.
➢ A through-beam sensor counts the resistors as they pass by.
➢ A counter preset for 20 (the amount of resistors in each kit) will automatically stop the take-up spool when the accumulated count equals 20.
➢ A second counter is provided to count the grand total.
➢ Manual reset buttons are provided for resetting both counters.

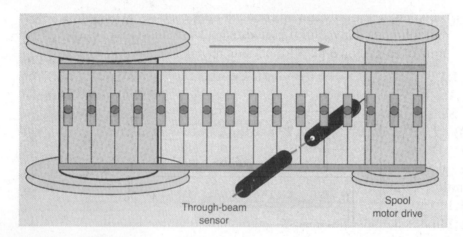

Through-beam sensor

Spool motor drive

Use the Silo Simulator screen and the following addresses to simulate the program:

Spool Drive Start Button _ I:1/0
Spool Drive Stop Button _ I:1/1
Through-Beam Sensor _ I:1/3
Manual Reset Kit _ I:1/6 (Pos B)
Manual Reset Total _ I:1/7 (Pos C)
Spool Motor _ O:2/0
CTU (Kit) _ C5:1
CTU (Total) _ C5:2

8-26 Write a documented PLC program that will latch on a light 20 seconds after an input switch has been turned on. The timer will continue to cycle up to 20 seconds and reset itself until the input switch has been turned off. After the third time the timer has timed to 20 seconds, the light will be unlatched. Use the I/O Simulator screen and the following addresses to simulate the program:

Input Switch _ I:1/0
Light _ O:2/0
TON _ T4:10
CTU _ C5:7

8-27 Write a documented PLC program that will turn a light on when a count reaches 20. The light is then to go off when a count of 30 is reached. Use the I/O Simulator screen and the following addresses to simulate the program:

Count Input _ I:1/0
Reset Input _ I:1/1
Light _ O:2/0
CTU _ C5:1
CTU _ C5:2

8-28 Write a documented PLC program that will implement the box-stacking process illustrated. This application requires the control of a conveyor belt that feeds a mechanical stacker. The stacker can stack various numbers of cartons of ceiling tile onto each pallet (depending on the pallet size and the preset value of the counter). When the required number of cartons has been stacked, the conveyor is stopped until the loaded pallet is removed and an empty pallet is placed onto the loading area. A photoelectric sensor will be used to provide count pulses to the counter after each carton passes by. In addition to a conveyor motor start/stop station, a remote reset switch is provided to allow the operator to reset the system from the forklift after an empty pallet is placed onto the loading area. The operation of this system can be summarized as follows:

➤ Pressing the start button starts the conveyor.
➤ As each box passes the photoelectric sensor, a count is registered.
➤ When the preset value is reached (in this case 20), the conveyor belt turns off.
➤ The forklift operator removes the loaded pallet.
➤ After the empty pallet is in position, the forklift operator activates the remote reset switch, which then starts the whole cycle over again.

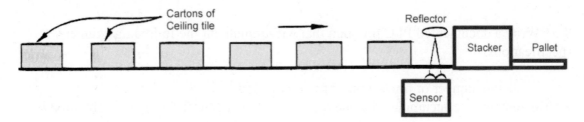

Use the Silo Simulator screen and the following addresses to simulate the program:

Conveyor Start Button _ I:1/0
Conveyor Stop Button _ I:1/1
Photoelectric Sensor _ I:1/3
Remote Reset Switch (NO) _ I:1/6 (Pos B)
Conveyor Motor _ O:2/0
Counter _ C5:1

8-29 Write a documented PLC program to operate a light according to the following sequence:

➢ A momentary pushbutton is pressed to start the sequence.
➢ The light is switched on and remains on for 2 seconds.
➢ The light is then switched off and remains off for 2 seconds.
➢ A counter is incremented by 1 after this sequence.
➢ The sequence then repeats for a total of four counts.
➢ After the fourth count, the sequence will stop and the counter will be reset.

Use the I/O Simulator screen and the following addresses to simulate the program:

Input PB (NO) _ I:1/0
Light _ O:2/0
CTU _ C5:1
TON _ T4:1
TON _ T4:2

8-30 Write a documented PLC program that will latch an output, PL1, after an input, PB1, has cycled on 20 times. When the count of 20 is reached, the counter will reset itself automatically. PB2 will unlatch PL1. Use the I/O Simulator screen and the following addresses to simulate the program:

PB1 (NO) _ I:1/0
PB2 (NO) _ I:1/1
PL1 _ O:2/0
CTU _ C5:1

8-31 Write a documented PLC program that will implement the following counter specifications:

➢ Counts the number of times a pushbutton is closed.
➢ Decrements the accumulated value of the counter each time a second pushbutton is closed.
➢ Turns on a pilot light, PL1, any time the accumulated value of the counter is less than 20.
➢ Turns on a second pilot light, PL2, when the accumulated value of the counter is equal to or greater than 20.
➢ Resets the counter to zero when a selector switch is closed.

Use the I/O Simulator screen and the following addresses to simulate the program:

Increment Count Button (NO) _ I:1/0
Decrement Count Button (NO) _ I:1/1
Reset Selector Switch _ I:1/2
PL1 _ O:2/0
PL2 _ O:2/1
Counter _ C5:1

8-32 Write a documented PLC program that will implement the following packaging process:

➢ The purpose of this process is to deposit 10 pieces of the product in each container.
➢ The process is set in operation by pressing a start button.
➢ As the product passes through the light beam, it is detected by the photoelectric proximity switch and counted by the PLC counter.
➢ When the count reaches 10, the solenoid-operated deflector plate (SOL A) energizes to channel the product from chute A to chute B.
➢ When the second count of 10 is reached, the solenoid-operated defector plate de-energizes to channel the product back to chute A, and so on.
➢ Provisions are also made for stopping the process at any time and manually resetting the accumulated value of the counters.

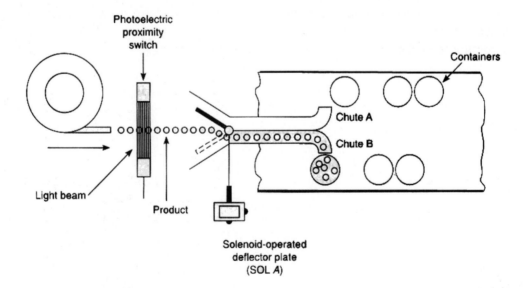

Use the Bottle Line Simulator screen and the following addresses to simulate the program:

Stop PB _ I:1/0
Start PB _ I:1/1
Proximity Sensor (NO) _ I:1/6
Reset (NO) _ I:1/3 (Selector Switch Pos. B)
Conveyor Motor M1 _ O:2/0
Conveyor Motor M2 _ O:2/2
Gate Solenoid A _ O:2/5
CTU _ C5:1
CTU _ C5:2

8-33 Write a documented PLC program that will simulate the operation of the hardwired relay control circuit shown. The sequence of operation is as follows:

➢ An operator initiates the system, and the items to be sorted are then fed onto the production line conveyor.
➢ Once on the conveyor, the items proceed, operating limit switch 1, which counts all items.
➢ Limit switch 2 counts only the larger items.
➢ As limit switch 2 is operated, a pneumatic ram is activated and thus stores all the larger items in packing box 2.
➢ The smaller items continue to the end of the conveyor and are stored in packing box 1.
➢ Actuating the reset switch at any time resets both counters.

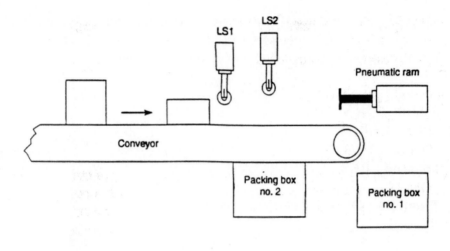

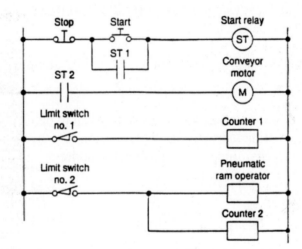

Use the Bottle Line Simulator screen and the following addresses to simulate the program:

Stop PB _ I:1/0
Start PB _ _ I:1/1
LS1 _ I:1/6
LS2 _ I:1/7
Reset (NO) _ I:1/3 (Selector Switch Pos. B)
Conveyor Motor M1 _ O:2/0
Conveyor Motor M2 _ O:2/2
Pneumatic Ram _ O:2/5
Counter 1 _ C5:1
Counter 2 _ C5:2

8-34(a) Using the Batch Simulator screen, write a program to determine the number of pulses that are generated by flowmeter FL1 in filling the tank from the low level to the high level. Use the following addresses in developing the program:

Fill Button (Start NO) _ I:1/0
Empty Button (Stop NC) _ I:1/1
Flowmeter FL1 _ I:1/5
Low-Level Sensor _ I:1/3
High-Level Sensor _ I:1/4
Pump P1 _ O:2/1
Discharge Pump P3 _ O:2/3
CTU _ C5:1

8-34(b) Repeat the original test using pump P2 **(O:2/2)** and flowmeter FL2 **(I:1/6)** in place of pump P1 and flowmeter FL1.

8-34(c) Repeat the original test using both pumps P1 and P2 along with an additional counter **C5:2.**

8-35(a) Using the Batch Simulator screen, write a program to determine the time it takes for pump PL1 (running continuously) to fill the tank from low level to high level. Also, determine the total gallons of liquid. Use the following addresses in developing the program:

Fill Button (Start NO) _ I:1/0
Empty Button (Stop NC) _ I:1/1
Flowmeter FL1 _ I:1/5
Low-Level Sensor _ I:1/3
High-Level Sensor _ I:1/4
Pump P1 _ O:2/1
Discharge Pump P3 _ O:2/3
CTU _ C5:1
RTO _ T4:1

8-35(b) Modify the program to determine the time it takes for pump P3 (running continuously) to empty the tank from the high level to the low level. Also, determine the flow rate in gallons per second.

8-36 Using the Batch Simulator screen, write a program that will implement the following tank filling and mixing sequence.

➢ The start button is pressed to start input pump P1.
➢ After 100 gallons (one flowmeter pulse per gallon) input pump P1 stops automatically and the mixer motor starts.

➢ The mixer motor is automatically stopped after 30 seconds, and the input pump P1 is started.
➢ After an additional 60 gallons are pumped into the tank, the input pump stops and the mixer motor starts.
➢ The mixer motor is operated for 20 seconds and stops.
➢ The discharge pump P3 motor is started automatically and runs until the low-level sensor is actuated.
➢ The process stops.

Use the following addresses in developing the program:

Start PB _ I:1/0
Stop PB _ I:1/1
Input Flowmeter _ I:1/5
Low-Level Sensor _ I:1/3
Mixer Motor _ O:2/0
Input Pump P1 _ O:2/1
Discharge Pump P3 _ O:2/3
Internal Relay _ B3:0/0
CTU _ C5:1
CTU _ C5:2
RTO _ T4:1
RTO _ T4:2

8-37 Using the Batch Simulator screen, write a program that will alternate the use of two pumps so that they both get the same amount of usage over their lifetime. The operation of the process can be summarized as follows:

➢ A single start/stop pushbutton station is provided for control of two input pumps P1 and P2.
➢ Initially, the start/stop pushbutton station is operated to control pump P1.
➢ When the tank is full, drain pump motor P3 is started automatically and runs until the low-level sensor is actuated.
➢ After five fillings of the tank by pump P1, control automatically shifts to pump P2.
➢ Operation of the start/stop pushbutton station now controls pump P2.
➢ After five fillings of the tank by pump P2, the sequence is repeated.

Use the following addresses in developing the program:

Start PB _ I:1/0
Stop PB _ I:1/1
Low-Level Sensor _ I:1/3
High-Level Sensor _ I:1/4
Input Pump P1 _ O:2/1
Input Pump P2 _ O:2/2
Discharge Pump P3 _ O:2/3
Internal Relay _ B3:0/0
CTU _ C5:1
CTU _ C5:2

8-38 Two ingredients, A and B, are to be heated and mixed. Two pumps, P1 and P2, provide the necessary pressure to send the ingredients through the lines. The procedure is as follows:

➢ The start pushbutton is pressed to start the process and pump P1.
➢ After 100 gallons of ingredient A are pumped into the tank, pump P1 stops automatically and the heater is automatically turned on.
➢ The heater is automatically turned off after 2 seconds, and pump P2 is energized.
➢ Ingredient B is pumped into the tank, by pump P2, until the liquid reaches the high-level sensor.
➢ The mixer motor starts automatically and runs for 10 seconds.
➢ After the mixer motor stops, P3 discharge pump starts automatically and runs until the low-level sensor is actuated.
➢ The process stops.
➢ The process can be stopped at any time, by means of the stop button, without loss of the counter accumulated count or the timer accumulated time.
➢ The idle light is energized whenever the process is stopped.
➢ The run light is energized whenever the process is in operation.
➢ The full light is energized whenever the tank is full.

Use the Batch Simulator screen and the following addresses in developing the program:

Start PB _ I:1/0
Stop PB _ I:1/1
Low-Level Sensor _ I:1/3
High-Level Sensor _ I:1/4
Ingredient A Flowmeter _ I:1/5
Input Pump P1 _ O:2/1
Input Pump P2 _ O:2/2
Discharge Pump P3 _ O:2/3
Mixer Motor _ O:2/0
Heater _ O:2/4
Idle PL _ O:2/6
Run PL _ O:2/5
Full PL _ O:2/7
Internal Relay _ B3:0/0
CTU _ C5:1
RTO _ T4:1
RTO _ T4:2

8-39 Implement the count-down counter program shown with the preset value set to 2 and the initial accumulated value set to 4. Use the I/O Simulator screen and the following addresses to simulate the program.

Count PB _ I:1/0
Count-Down-Counter _ C5:0
Output PL _ O:2/0

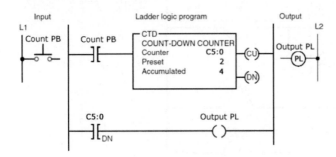

CHAPTER **9**

Program Control Instructions

LogixPro Programming Assignments

9-1 Implement the master control reset (MCR) program shown. Demonstrate that when the MCR instruction is de-energized, all nonretentive outputs de-energize and all retentive outputs remain in their last state. Use the I/O Simulator screen and the following addresses to simulate the program:

ON/OFF Switch _ I:1/0
Stop PB _ I:1/1
Start PB _ I:1/2
Level Switch _ I:1/3
LS1 _ I:1/4
LS2 _ I:1/5
M _ O:2/0
PL1 _ O:2/1
SOL _ O:2/2

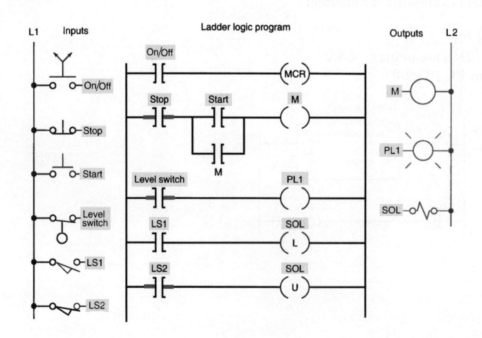

9-2 Implement the MCR fenced zone PLC program shown. Demonstrate how the rungs between the two MCR instructions are controlled. Use the I/O Simulator screen and the following addresses to simulate the program:

Input A _ I:1/0
Input B _ I:1/1
Input C _ I:1/2
Input D _ I:1/3
Input E _ I:1/4
Output A _ O:2/0
Output B _ O:2/1

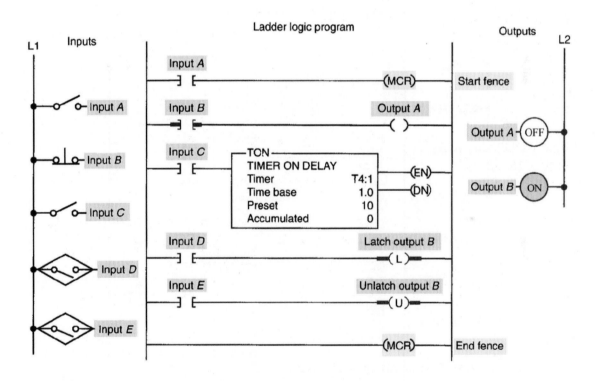

9-3 Implement the jump-to-label PLC program shown. Demonstrate how the jump-to-label instruction is executed. Use the I/O Simulator screen and the following addresses to simulate the program:

PB1 _ I:1/0
PB2 _ I:1/1
PS1 _ I:1/2
LLS1 _ I:1/3
LS1 _ I:1/4
LS2 _ I:1/5
LS3 _ I:1/6

PB3 _ I:1/7
LS4 _ I:1/8
TS1 _ I:1/9
M _ O:2/0
PL1 _ O:2/1
SOL 1 _ O:2/2
SOL 2 _ O:2/3
PL2 _ O:2/4
SOL 3 _ O:2/5
SOL 4 _ O:2/6
Heater _ O:2/7

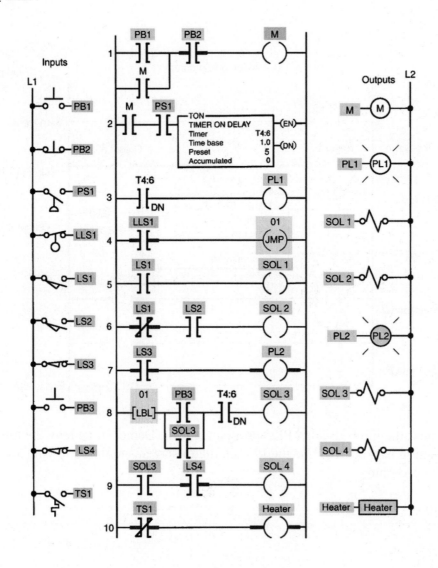

9-4 Implement the jump-to-subroutine PLC program shown. Demonstrate how the jump-to-subroutine instruction is executed. Use the I/O Simulator screen and the following addresses to simulate the program:

OFF/ON _ **I:1/0**
Stop _ I:1/1
Start _ I:1/2
Sensor _ I:1/3
Motor (M1) _ O:2/0
PL1 _ O:2/1
SOL _ O:2/2
Timers _ T4:0 & T4:1

Main Program
File 2

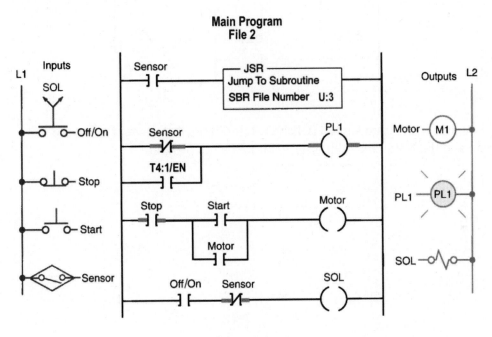

Subroutine
File 3

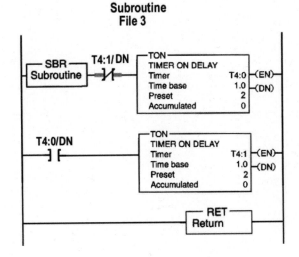

9-5 Implement the PLC program shown using the I/O Simulator screen and the addresses given. Demonstrate how each of the following is executed:

➢ Forcing input I:1/3 on and off.
➢ Forcing output O:2/5 on and off.
➢ Forcing output O:2/6 on and off.

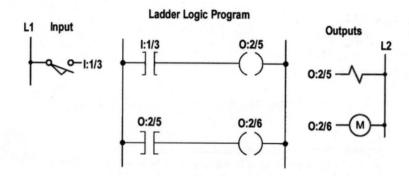

9-6 Implement the MCR program shown (Chapter 9, problem 1 of the text). Use the I/O Simulator screen and the following addresses to simulate the program:

S1 _ I:1/1
S2 _ I:1/2
S3 _ I:1/3
S4 _ I:1/4
S5 _ I:1/5
S6 _ I:1/6
PL1 _ O:2/1
PL2 _ O:2/2
TON _ T4:1

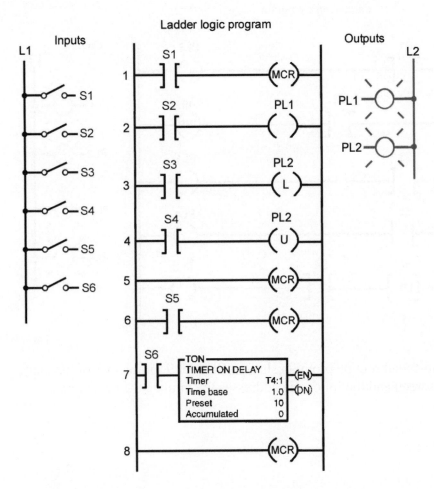

Ladder logic program

9-7 Implement the jump-to-label program shown (Chapter 9, problem 2 of the text). Use the I/O Simulator screen and the following addresses to simulate the program:

S1 _ I:1/1
S2 _ I:1/2
S3 _ I:1/3
S4 _ I:1/4
S5 _ I:1/5
PL1 _ O:2/1
PL2 _ O:2/2
PL3 _ O:2/3
PL4 _ O:2/4

Ladder logic program

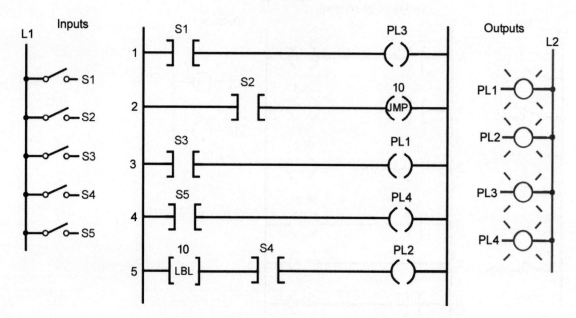

9-8 Implement the jump-to-subroutine program shown (Chapter 9, problem 3 of the text). Use the I/O Simulator screen and the following addresses to simulate the program:

S1 _ I:1/1
S2 _ I:1/2
S3 _ I:1/3
S4 _ I:1/4
S5 _ I:1/5
PL1 _ O:2/1
PL2 _ O:2/2
PL3 _ O:2/3
PL4 _ O:2/4

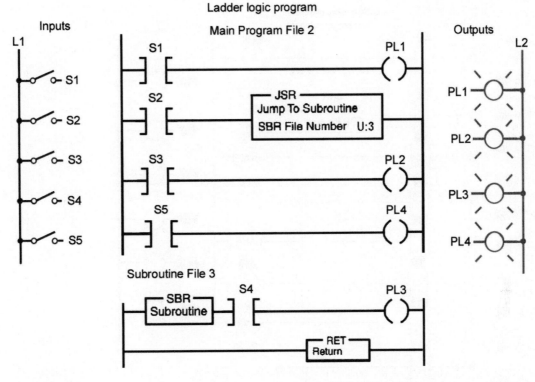

Ladder logic program

9-9 Implement the jump-to-subroutine program shown (Chapter 9, problem 4 of the text). Use the I/O Simulator screen and the following addresses to simulate the program:

S1 _ I:1/1
S2 _ I:1/2
S3 _ I:1/3
S4 _ I:1/4
S5 _ I:1/5
S6 _ I:1/6
S7 _ I:1/7
S8 _ I:1/8
S9 _ I:1/9
S10 _ I:1/10
S11 _ I:1/11
S12 _ I:1/12
S13 _ I:1/13
PL1 _ O:2/1
PL2 _ O:2/2
PL3 _ O:2/3
PL4 _ O:2/4
PL5 _ O:2/5
PL6 _ O:2/6

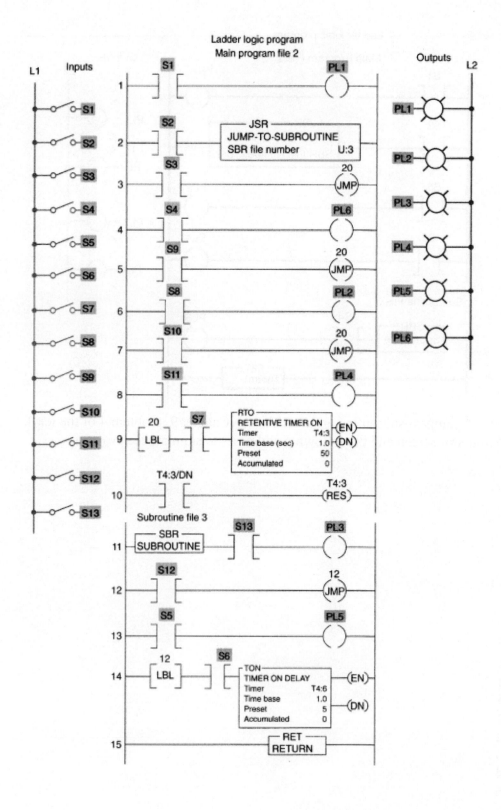

Ladder logic program
Main program file 2

9-10 Implement the MCR program shown. Use the I/O Simulator screen and the following addresses to simulate the program:

S1 _ I:1/1
S2 _ I:1/2
S3 _ I:1/3
S4 _ I:1/4
S5 _ I:1/5
S6 _ I:1/6
S7 _ I:1/7
PL1 _ O:2/1
PL2 _ O:2/2
PL3 _ O:2/3

Operate the program according to the following sequence, and answer the question(s) associated with each sequence:

(a) Close switches 1, 2, 3, 4, and 6, and allow timer T4:2 to time out. What lights are on?
(b) Open switch 1. What light is on now? Why did pilot lights PL1 and PL2 go off?
(c) Open switch 4 and close switch 5. Did pilot light PL1 go off? Why or why not?
(d) What happened to the two timers when you disabled the MCR zone?
(e) What happened to the two timers when you reenabled the MCR zone?

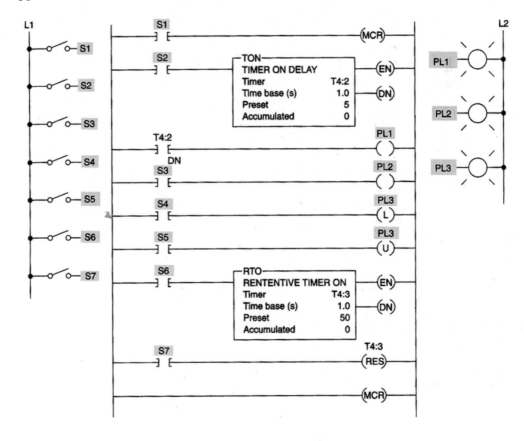

9-11 Construct the example of the subroutine (SBR) and the return instruction (RET) program shown. The purpose of this program is to find the average of the value stored in N7:5 and N7:20 and store the result in N7:30. This occurs only when input S1 is true and is accomplished by passing parameters to the subroutine and doing the math in the subroutine and then returning the answer to the main program through the RET instruction. Prove the operation by using the data monitor to insert values for N7:5 and N7:20, and verify that the average value is contained in N7:30.

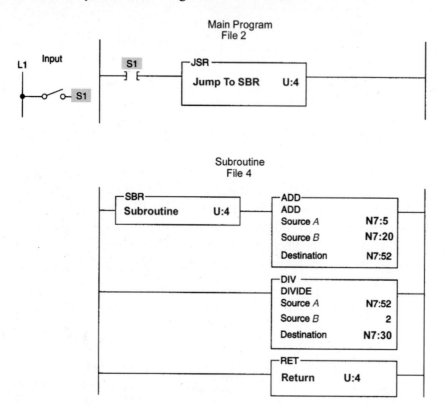

9-12 Construct the example of the temporary end (TND) program shown. Demonstrate how this instruction can be used to progressively debug the program. Use the I/O Simulator screen and the following addresses to simulate the program:

S1 _ I:1/0
S2 _ I:1/1
S3 _ I:1/2
S4 _ I:1/3
S5 _ I:1/4
S6 _ I:1/5
S7 _ I:1/6
PL1 _ O:2/0
PL2 _ O:2/1

PL3 _ O:2/2
PL4 _ O:2/3
PL5 _ O:2/4
PL6 _ O:2/5

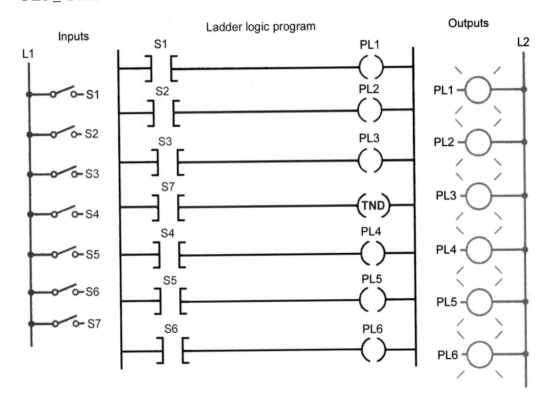

9-13 Construct the example of the jump (JMP) program shown. Use the I/O Simulator screen and the following addresses to simulate the program:

PB _ I:1/1
SW _ I:1/0
PL1 _ O:2/1
PL2 _ O:2/2
JMP and LBL _ 1
Internal Relay _ B3:0/1

Answer the following with regards to the operation of the program:

- When the switch is open, pressing the pushbutton activates which pilot light(s)? Why?
- When the switch is closed, pressing the pushbutton activates which pilot light(s)? Why?

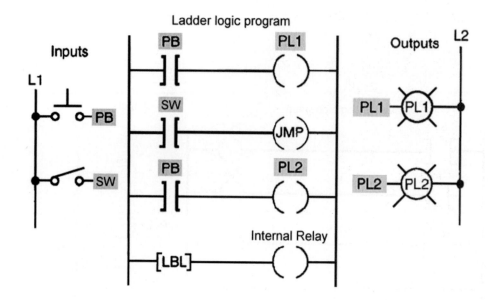

CHAPTER **10**

Data Manipulation Instructions

LogixPro Programming Assignments

10-1 Implement the Move (MOV) instruction program shown. Prove the operation by using the data monitor to insert a value for N7:30, and verify that this value is moved to N7:20 when input A **(I:1/0)** is true. Use the I/O Simulator screen to simulate the program.

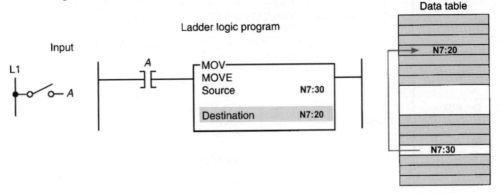

10-2 Implement the Masked Move (MVM) instruction program shown. Prove the operation by using the data monitor to insert the given values for B3:0, B3:1, and B3:4 prior to input A **(I:1/0)** going true. Verify that the masked value is moved to B3:4 when input A goes true. Use the I/O Simulator screen to simulate the program.

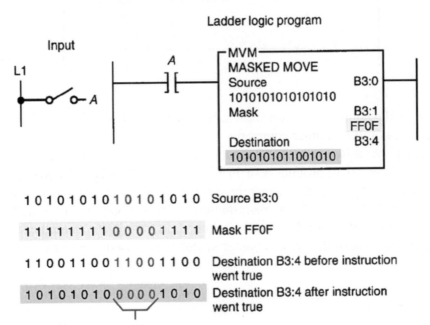

10-3 Implement the changing of the preset value of a timer using the Move instruction program shown. Use the I/O Simulator screen and the following addresses to simulate the program:

PB1 _ I:1/0
SS1 _ I:1/1
PL1 _ O:2/0
TON _ T4:1

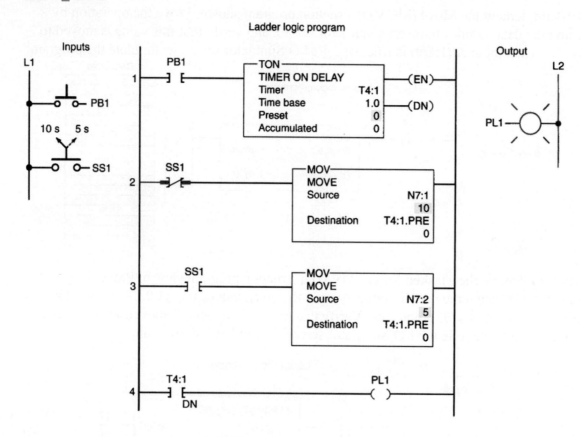

Ladder logic program

10-4 Implement the counter program shown. In this application, a limit switch programmed to operate a counter counts the products coming off of a conveyor line onto a storage rack. Three different types of products are run on this line. The storage rack has room for only 300 boxes of product A or 175 boxes of product B or 50 boxes of product C. Three switches are provided to select the desired preset counter value depending on the product line—A, B, or C—being manufactured. A reset button is provided to reset the accumulated count to zero. A pilot light is switched on to indicate when the storage rack is full. If more than one of the preset counter switches is closed, the last value is selected.

Use the I/O Simulator screen and the following addresses to simulate the program:

LS1 _ I:1/0
A _ I:1/1
B _ I:1/2
C _ I:1/3
Reset _ I:1/4
Full PL _ O:2/0
CTU _ C5:1

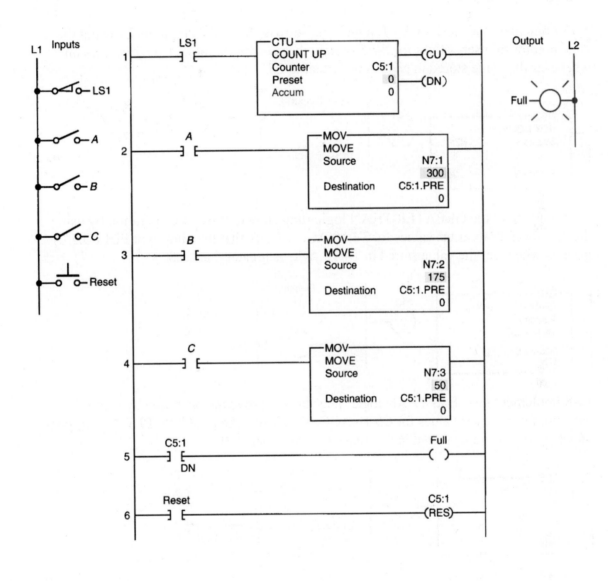

10-5 Implement the EQUAL logic rung shown. Prove the operation by using the data monitor to insert values for T4:0.ACC and N7:40. Verify that the pilot light PL1 (O:2/0) goes on whenever these two values are equal.

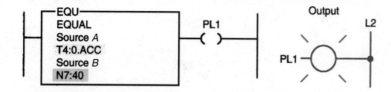

10-6 Implement the NOT EQUAL logic rung shown. Prove the operation by using the data monitor to insert values for N7:5. Verify that the pilot light PL1 (O:2/0) goes on whenever the value stored in N7:5 is not equal to 25.

10-7 Implement the GREATER THAN logic rung shown. Prove the operation by using the data monitor to insert values for T4:10.ACC. Verify that the pilot light PL1 (O:2/0) goes on whenever the value stored in T4:10.ACC is greater than 200.

10-8 Implement the LESS THAN logic rung shown. Prove the operation by using the data monitor to insert values for C5:10.ACC. Verify that the pilot light PL1 (O:2/0) goes on whenever the value stored in C5:10.ACC is less than 350.

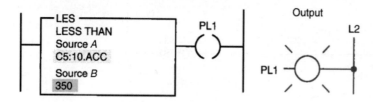

10-9 Implement the GREATER THAN OR EQUAL logic rung shown. Prove the operation by using the data monitor to insert values for N7:55 and N7:12. Verify that the pilot light PL1 (O:2/0) goes on whenever the value stored in N7:55 is greater than or equal to that stored in N7:12.

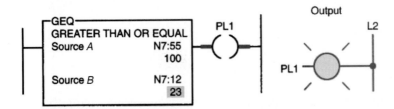

10-10 Implement the LESS THAN OR EQUAL logic rung shown. Prove the operation by using the data monitor to insert values for C5:1.ACC. Verify that the pilot light PL1 (O:2/0) goes on whenever the value stored in C5:1.ACC is less than or equal to 457.

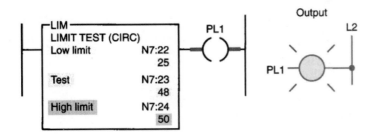

10-11 Implement the LIMIT TEST logic rung shown. Prove the operation by using the data monitor to insert test values (N7:23). Verify that the pilot light PL1 (O:2/0) goes on whenever the test value stored in N7:23 ranges from 25 to 50.

10-12 Implement the LIMIT TEST logic rung shown. Prove the operation by using the data monitor to insert test values (N7:29). Verify that the pilot light PL1 (O:2/0) goes on whenever the test value stored in N7:29 is less than 50 or greater than 100.

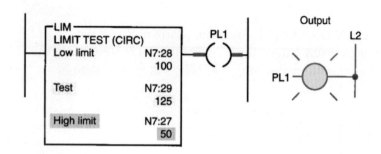

10-13 Implement the masked comparison for equal (MEQ) logic rung shown using the I/O Simulator screen. This example masks out the upper three thumbwheel switches of (I:5) in the I/O Simulator and energizes pilot light PL1 (O:2/0) if the lowest thumbwheel is equal to 9. Verify the operation by changing the values of the thumbwheel switches.

Ladder logic program

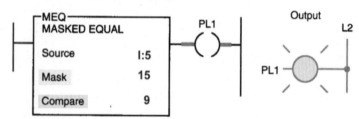

10-14 Write a documented program for the relay schematic shown using only one internal timer along with data compare statements. Use the I/O Simulator screen and the following addresses to simulate the program:

Stop PB _ I:1/0
Start PB _ I:1/1
SOL A _ O:2/0
SOL B _ O:2/1
SOL C _ O:2/2
SOL D _ O:2/3
TON _ T4:1

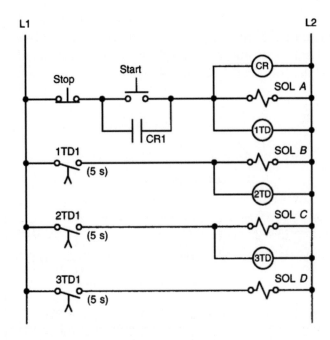

10-15 Implement the on delay data compare program shown. Demonstrate that when the switch is closed, the light comes on after 5 seconds, stays on for 10 seconds, and then turns off. Use the I/O Simulator screen and the following addresses to simulate the program:

S1 _ I:1/0
PL1 _ O:2/0
TON _ T4:1

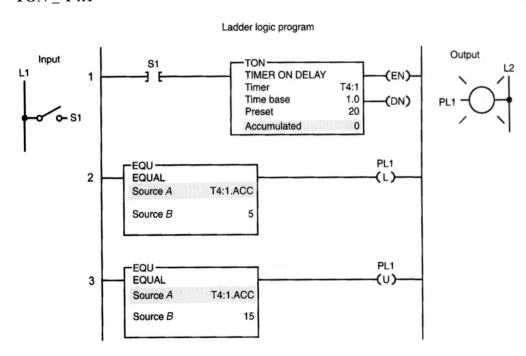

10-16 Implement the counter data compare program shown. Demonstrate that the output will be energized when the accumulated value of the counter is between 0 and 19 and that the counter will reset automatically when it reaches its preset value of 50. Use the I/O Simulator screen and the following addresses to simulate the program:

Count PB _ I:1/0
PL1 _ O:2/0
CTU _ C5:1

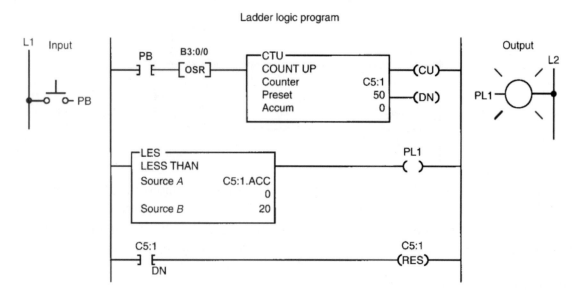

Ladder logic program

10-17 Implement the word level thumbwheel switch and LED data compare program shown. Demonstrate that the decimal setting of the thumbwheel switches is monitored by the LED display board. Verify that pilot light is energized whenever S1 is closed and the value of the thumbwheel switch setting equals 10 Hex-BCD (which is 0001 0000 binary and 16 decimal). Use the BCD Simulator that is part of the I/O Simulator screen and the following addresses to simulate the program:

S1 _ I:1/0
TWS _ I:5 (word level)
PL _ O:2/0
LED Display _ O:6 (word level)

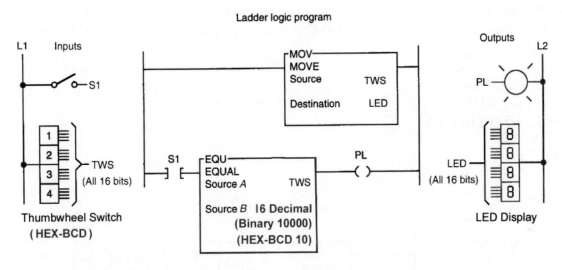

10-18 Implement the set-point temperature control program shown. The tank is to maintain a temperature of 102 (Hex-BCD) with a variation from 100 (Hex-BCD) to 104 (Hex-BCD) between the ON and OFF cycles. Temperature control is to be applied any time the process is running. Demonstrate the correct operation of the program. Use the Batch Simulator screen and the following addresses to simulate the program:

S1 (ON/OFF) _ I:1/10 (Pos B)
Thermocouple _ I:3 (vary thumbwheel switch setting for temperature)
Heater _ O:2/4
LED Display _ O:4 (word level)

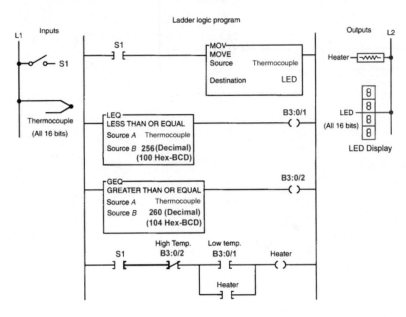

10-19 Implement the data transfer counter program shown (Chapter 10, problem 2). Use the BCD Simulator that is part of the I/O Simulator screen and the following addresses to simulate the program:

Count PB _ I:1/0
Reset PB _ I:1/1
TWS _ I:5 (word level)
PL _ O:2/0
CTU _ C5:1

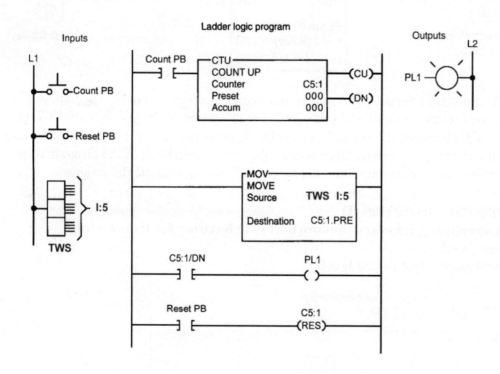

10-20 Implement the data compare program shown (Chapter 10, problem 4). Use the BCD Simulator that is part of the I/O Simulator screen and the following addresses to simulate the program:

S1 _ I:1/0
TWS _ I:5 (word level)
PL _ O:2/0

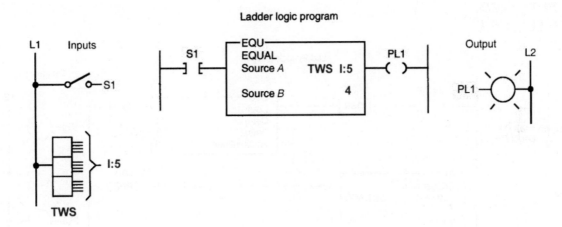

10-21 Implement the data compare program shown (Chapter 10, problem 5). Use the BCD Simulator that is part of the I/O Simulator screen and the following addresses to simulate the program:

S1 _ I:1/0
TWS _ I:5 (word level)
PL1 _ O:2/0

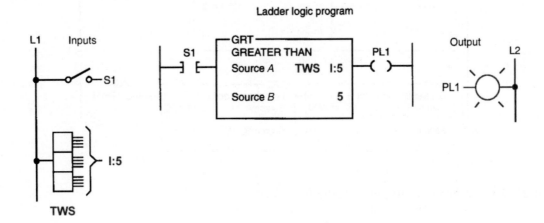

10-22 Implement the data compare program shown. Determine at what accumulated counter values pilot lights PL1, PL2, and PL3 are energized and the highest count before the counter is reset. Use the I/O Simulator screen and the following addresses to simulate the program:

Count PB _ I:1/0
PL1 _ O:2/0
PL2 _ O:2/1
PL3 _ O:2/2
CTU _ C5:1

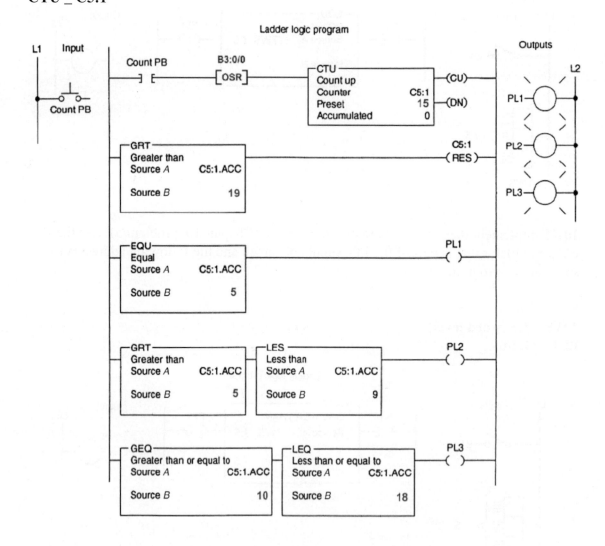

Ladder logic program

10-23 Write a program to perform the following:

➢ Turn on pilot light 1 (PL1) if the thumbwheel switch value is less than 4.
➢ Turn on pilot light 2 (PL2) if the thumbwheel switch value is equal to 4.
➢ Turn on pilot light 3 (PL3) if the thumbwheel switch value is greater than 4.

➢ Turn on pilot light 4 (PL4) if the thumbwheel switch value is less than or equal to 4.
➢ Turn on pilot light 5 (PL5) if the thumbwheel switch value is greater than or equal to 4.

Use the BCD Simulator that is part of the I/O Simulator screen and the following addresses to simulate the program:

TWS _ I:5 (word level)
PL1 _ O:2/1
PL2 _ O:2/2
PL3 _ O:2/3
PL4 _ O:2/4
PL5 _ O:2/5

10-24 Write a program that uses the mask move instruction to move only the upper 8 bits of the values stored at address I:5 to address O:6 and ignore the lower 8 bits. Use the BCD Simulator that is part of the I/O Simulator screen to simulate the program.

10-25 Write a program that will cause a pilot light to come on only if a PLC counter has a value of 6 or 10. Use the I/O Simulator screen and the following addresses to simulate the program:

Count PB (NO) _ I:1/0
Reset PB (NO) _ I:1/1
Pilot Light _ O:2/0
CTU _ C5:1

10-26 Using the LIM instruction, write a program that will cause a pilot light to come on if a PLC counter value is less than 10 or more than 30. Use the I/O Simulator screen and the following addresses to simulate the program:

Count PB (NO) _ I:1/0
Reset PB (NO) _ I:1/1
Pilot Light _ O:2/0
CTU _ C5:1

10-27 Using compare instructions, write a counter program to perform the following:

➢ Turn on PL1 when the count is 5.
➢ Turn on PL2 when the count is 10—at which time PL1 should turn off.
➢ Turn on PL3 when the count is 15—at which time PL2 should turn off.
➢ Reset the counter when the count is 20—at which time PL3 should switch off.

Use the I/O Simulator screen and the following addresses to simulate the program:

Count PB (NO) _ I:1/0
PL1 _ O:2/0
PL2 _ O:2/1
PL3 _ O:2/2
CTU _ C5:1

10-28 Using compare instructions, write a program that uses a single timer to turn pilot lights 1 through 8 on in sequence at 1-second intervals. When all eight lights are on, they are then turned off from 8 to 1 in sequence at 1-second intervals. The pattern is repeated as long as the ON/OFF switch is closed.

Use the I/O Simulator screen and the following addresses to simulate the program:

ON/OFF Switch _ I:1/0
PL1 _ O:2/0
PL2 _ O:2/1
PL3 _ O:2/2
PL4 _ O:2/3
PL5 _ O:2/4
PL6 _ O:2/5
PL7 _ O:2/6
PL8 _ O:2/7
TON _ T4:1

10-29 Using compare instructions, write a program to perform the following:

➤ Turn on PL1 (O:2/0) if the value of input word I:1 is 50.
➤ Turn on PL2 (O:2/1) if the value of input word I:1 is 56 .
➤ Turn on PL3 (O:2/2) if the value of input word I:1 is not equal to 176.
➤ Turn on PL4 (O:2/3) if the value of input word I:1 is 48, 49, 50, 51, or 52.
➤ Turn on PL5 (O:2/4) if the value of input word I:1 is 16 or 240.
➤ Turn on PL6 (O:2/5) only if NO inputs I:1/4, I:1/0, and I:1/7 are closed and all others are open.
➤ Turn on PL7 (O:2/6) only if all NO inputs except I:1/5 are open.
➤ Turn on PL8 (O:2/7) only if all NO inputs except I:1/3 are closed.

Use the compare instructions and the I/O Simulator screen to prove the operation of each rung.

10-30 Using a single timer and data compare instructions, develop a program that will operate cylinders in the desired sequence. The time between each step is to be 3 seconds. The desired sequence of operation will be as follows:

➢ All cylinders to retract.
➢ Cylinder 1 advance.
➢ Cylinder 1 retract and cylinder 3 advance.
➢ Cylinder 2 advance and cylinder 5 advance.
➢ Cylinder 4 advance and cylinder 2 retract.
➢ Cylinder 3 retract and cylinder 5 retract.
➢ Cylinder 6 advance and cylinder 4 retract.
➢ Cylinder 6 retract.
➢ Sequence to repeat.

Use the I/O Simulator screen and the following addresses to simulate the program:

Sequence Stop PB (NC) _ I:1/0
Sequence Start PB (NO) _ I:1/1
Cylinder 1 _ O:2/0
Cylinder 2 _ O:2/1
Cylinder 3 _ O:2/2
Cylinder 4 _ O:2/3
Cylinder 5 _ O:2/4
Cylinder 6 _ O:2/5
Internal Relay _ B3:0
TON _ T4:1

10-31 Write a program that will implement the control of traffic lights in two directions using a single timer and data compare instructions according to the timing chart shown.

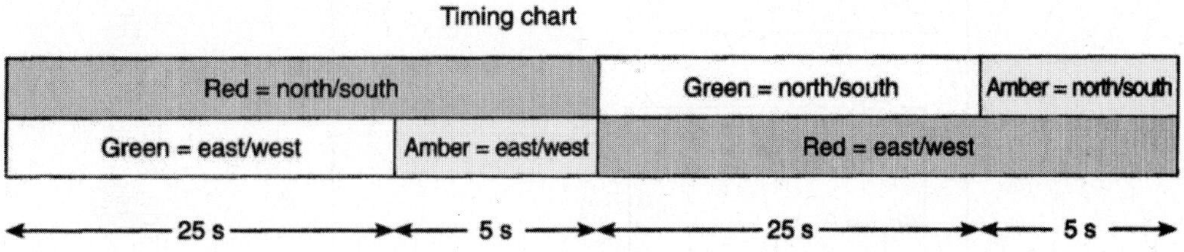

Timing chart

Red = north/south		Green = north/south	Amber = north/south
Green = east/west	Amber = east/west	Red = east/west	
←——— 25 s ———→	← 5 s →←	——— 25 s ———→	← 5 s →

Use the Traffic Simulator screen and the following addresses to simulate the program:

Red Light (N/S) _ O:2/0
Amber Light (N/S) _ O:2/1
Green Light (N/S) _ O:2/2
Red Light (E/W) _ O:2/4

Amber Light (E/W) _ O:2/5
Green Light (E/W) _ O:2/6
TON _ T4:1

10-32 Implement the data compare program shown. Determine at what accumulated counter values pilot lights LESS, EQU, and GRT are energized and the highest count before the counter is reset. Use the I/O Simulator screen and the following addresses to simulate the program:

Counter Input _ I:1/0
Counter Reset _ I:1/1
LESS PL _ O:2/0
EQU PL _ O:2/1
GRT PL _ O:2/2
CTU _ C5:1

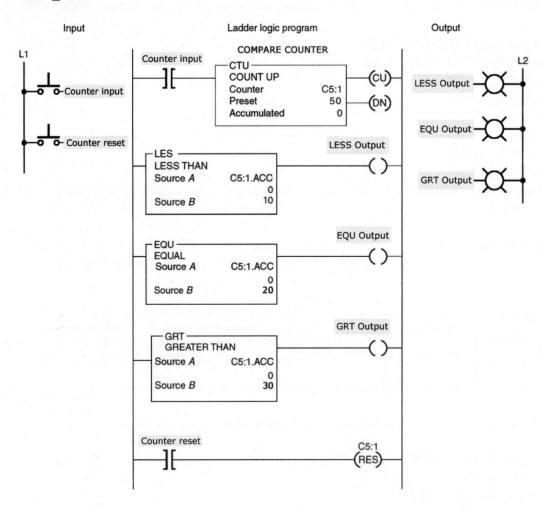

CHAPTER **11**

Math Instructions

LogixPro Programming Assignments

11-1 Implement the ADD instruction program shown. Prove the operation by using the data monitor to insert the given values for N7:0 and N7:1. Demonstrate that when input A **(I:1/0)** is true, the sum of these two numbers is stored at address N7:2. Use the I/O Simulator screen to simulate the program.

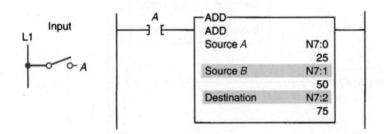

Ladder logic program

11-2 Implement the counter program shown that makes use of the ADD instruction. This application requires a light to come on when the sum of the counts from the two counters is equal to or greater than 350. Use the I/O Simulator screen and the following addresses to simulate the program:

LS1 _ I:1/0
LS2 _ I:1/1
Reset _ I:1/2
PL1 _ O:2/0
CTU _ C5:0
CTU _ C5:1

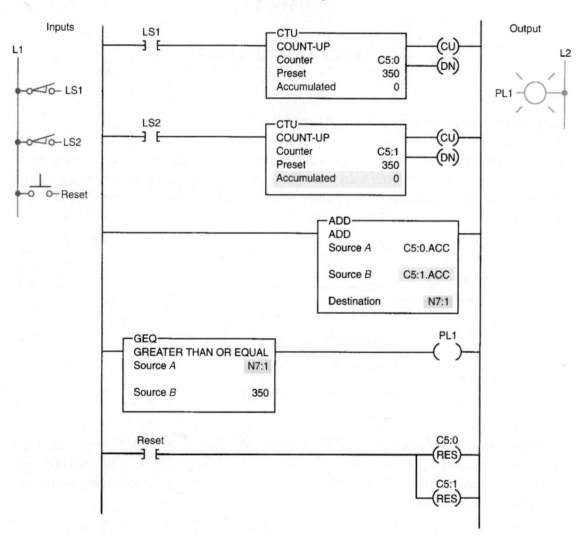

Ladder logic program

11-3 Implement the SUBTRACT instruction program shown. Prove the operation by using the data monitor to insert the given values for N7:10 and N7:05. Demonstrate that when input A **(I:1/0)** is true, the difference of these two numbers is stored at address N7:20. Use the I/O Simulator screen to simulate the program.

11-4 Implement the overfill alarm program shown. In this application the SUBTRACT function is used to indicate a vessel overflow condition. It requires an alarm to sound when the system leaks 2 lb or more of raw material into the vessel after a preset weight of 5 lb has been reached. Use the BCD Simulator that is part of the I/O Simulator screen and the following addresses to simulate the program:

Stop PB _ I:1/0
Start PB _ I:1/1
Weight Transducer _ I:5
Fill Solenoid _ O:2/0
Filling PL _ O:2/1
Full PL _ O:2/2
Alarm _ O:2/3

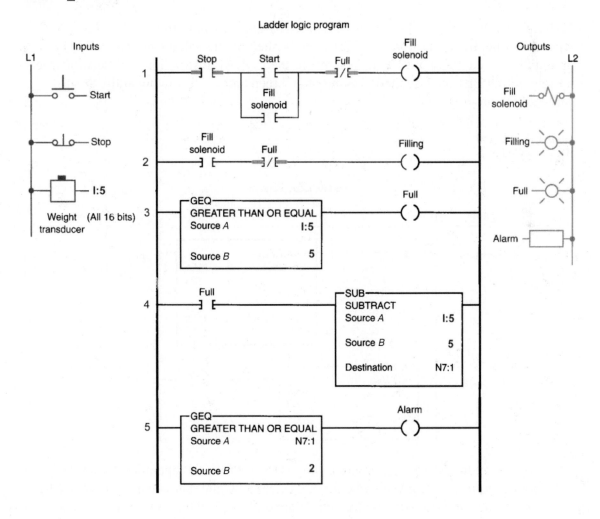

Ladder logic program

11-5 Implement the MULTIPLY instruction rung shown. Prove the operation by using the data monitor to insert the given value for N7:1. Demonstrate that when input A **(I:1/0)** is true, the product is stored at address N7:2. Use the I/O Simulator screen to simulate the program.

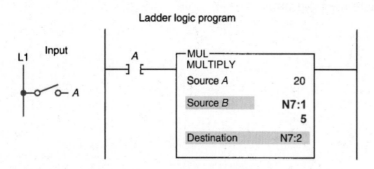

11-6 Implement the MULTIPLY instruction program shown. In this application when input A is true, the value stored at N7:1 is multiplied by the value stored at N7:2 and the product is placed into word N7:3. As a result, the EQUAL instruction will become true, turning output PL1 on. Use the I/O Simulator screen and the following addresses to simulate the program:

Input A _ I:1/0
PL1 _ O:2/0

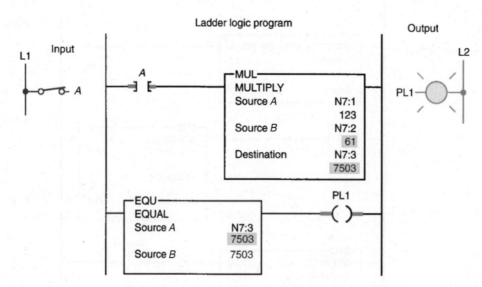

11-7 Implement the DIVIDE instruction rung shown. Prove the operation by using the data monitor to insert the given value for N7:1. Demonstrate that when input A **(I:1/0)** is true, the answer to the division is stored at address N7:3. Use the I/O Simulator screen to simulate the program.

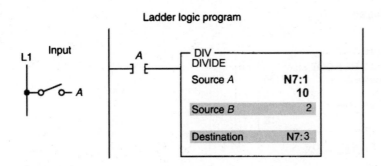

11-8 Implement the DIVIDE instruction program shown. In this application when input A is true, the value stored at N7:0 is divided by 4 and the answer is placed into word N7:5. As a result, the EQUAL instruction will become true, turning output PL1 on. Use the I/O Simulator screen and the following addresses to simulate the program:

Input A _ I:1/0
PL1 _ O:2/0

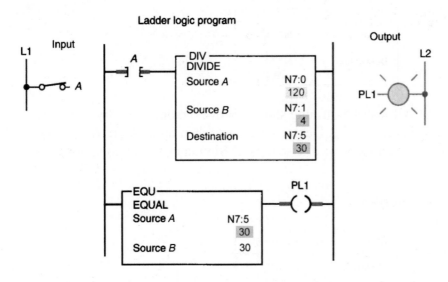

11-9 Implement the SQUARE ROOT instruction rung shown. Prove the operation by using the data monitor to insert the given value for N7:50. Demonstrate that when input A **(I:1/0)** is true, the square root of the number is stored at address N7:55. Use the I/O Simulator screen and the following address to simulate the program.

Input A _ I:1/0

Ladder logic program

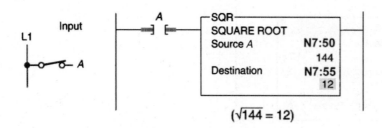

$$(\sqrt{144} = 12)$$

11-10 Implement the NEGATE instruction rung shown. Prove the operation by using the data monitor to insert the given value for N7:52. Demonstrate that when input A **(I:1/0)** is true, the sign of the value is changed and is stored at address N7:55. Use the I/O Simulator screen to simulate the program.

Ladder logic program

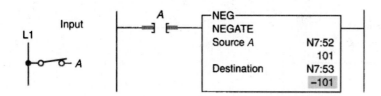

11-11 Implement the CLEAR instruction rung shown. Prove the operation by using the data monitor to insert a value for N7:22. Demonstrate that when input A **(I:1/0)** is true, the value stored in N7:22 is cleared to zero. Use the I/O Simulator screen to simulate the program.

Ladder logic program

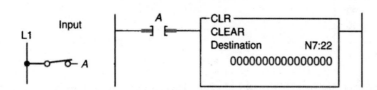

11-12 Implement the CONVERT TO BCD (TOD) instruction rung shown. This application converts the binary bit pattern at the source address, N7:23, into a BCD bit pattern in order to drive the LED display output **O:6,** which functions in BCD format. Use the BCD Simulator that is part of the I/O Simulator screen to simulate the program.

Ladder logic program

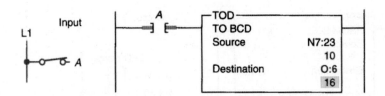

11-13 Implement the CONVERT FROM BCD (FRD) instruction rung shown. This application converts the BCD bit pattern stored at the source address, I:5, into a binary bit pattern of the same decimal value at the destination address N7:24. Use the BCD Simulator that is part of the I/O Simulator screen to simulate the program.

Ladder logic program

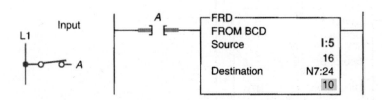

11-14 Implement the conversion from Celsius to Fahrenheit program shown. In this application, the thumbwheel switch indicates Celsius temperature. The program is designed to convert the Celsius temperatures to Fahrenheit values for display. Use the BCD Simulator that is part of the I/O Simulator screen to simulate the program.

Ladder logic program

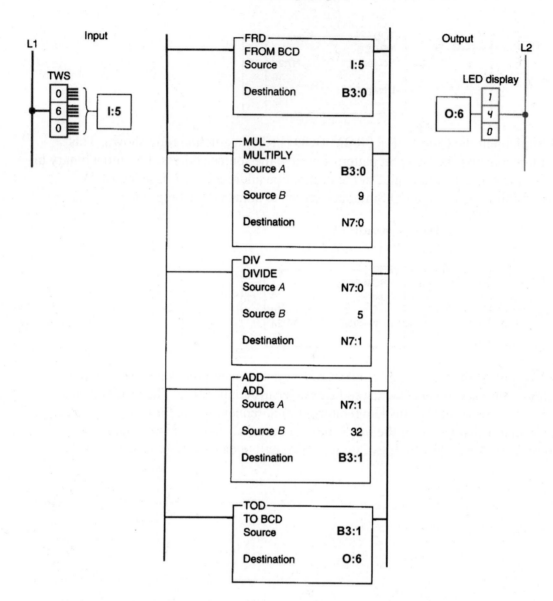

11-15 Design a program that will add the values stored at N7:23 and N7:24 and store the result in N7:30 whenever input A **(I:1/0)** is true, and then, when input B **(I:1/1)** is true, will copy the data from N7:30 to N7:31. Use the I/O Simulator screen to simulate the program.

11-16 Design a program that will take the accumulated value from TON timer **T4:1** and display it on a four-digit BCD format set of LEDs **(O:6)**. Include the provision to change the preset value of the timer from a set of four-digit BCD thumbwheels **(I:5)** when input A **(I:1/1)** is true. Use the BCD Simulator that is part of the I/O Simulator screen to simulate the program.

11-17 Design a program that will implement the following arithmetic operation whenever input A **(I:1/1)** is true:

➢ Use a move instruction to place the value 45 in N7:10 and 286 in N7:1.
➢ Add the values together and store the result in N7:2.
➢ Subtract the value in N7:2 from 785 and store the result in N7:3.
➢ Multiply the value in N7:3 by 25 and store the result in N7:4.
➢ Divide the value in N7:3 by 2 and store the result in N7:8.

Use the I/O Simulator screen to simulate the program.

11-18(a) There are three parts conveyor lines (1, 2, and 3) feeding a main conveyor. Each of the three conveyor lines has its own counter. Construct a PLC program to obtain the total count of parts on the main conveyor and display this number as a decimal value on a four-digit BCD format LED display. Use the I/O Simulator screen and the following addresses to simulate the program:

Line 1 Count Limit Switch _ I:1/0
Line 2 Count Limit Switch _ I:1/1
Line 3 Count Limit Switch _ I:1/2
Reset NO PB _ I:1/3
BCD Display _ O:6
Counters _ C5:0, C5:1, C5:2

11-18(b) Add a timer (T4:0) to the program that will update the total count every 3 seconds.

11-19 Two parts conveyor lines, A and B, feed a main conveyor line M. A third conveyor line, R, removes rejected parts a short distance away from the main conveyor. Conveyors A, B, and R have parts counters connected to them. Construct a PLC program to obtain the parts output of main conveyor M and display this number as a decimal value on a four-digit BCD format LED display. Use the I/O Simulator screen and the following addresses to simulate the program:

Line A Count Limit Switch _ I:1/0
Line B Count Limit Switch _ I:1/1
Line R Count Limit Switch _ I:1/2
Reset NO PB _ I:1/3
BCD Display _ O:6
Counters _ C5:1, C5:2, C5:3

11-20 A main conveyor has two conveyors, A and B, feeding it. Feeder conveyor A puts six packs of canned soda on the main conveyor. Feeder B puts eight packs of canned soda on the main conveyor. Both feeder conveyors have counters that count the number of packs leaving them. Construct a PLC program to give the total can count on the main conveyor and display this number as a decimal value on a four-digit BCD format LED display. Use the I/O Simulator screen and the following addresses to simulate the program:

Conveyor A Count Limit Switch _ I:1/0
Conveyor B Count Limit Switch _ I:1/1
Reset NO PB _ I:1/2
BCD Display _ O:6
Counters _ C5:1, C5:2

CHAPTER **12**

Sequencer and Shift Register Instructions

LogixPro Programming Assignments

12-1 Implement the SEQUENCER OUTPUT (SQO) instruction rung shown. This single instruction identifies where the output data are stored (file #B3:0), the destination of that output data (word O:2), and the length or number of steps (4) in the sequence. The mask is a filter through which all data from the sequencer file must pass before being placed in the output word. Only bits in the mask that are set (1) will pass data to the destination. This instruction also tracks what the current sequencer position is. Use the I/O Simulator screen and the following addresses to simulate the program:

PB1 _ I:1/0
Output Word _ O:2

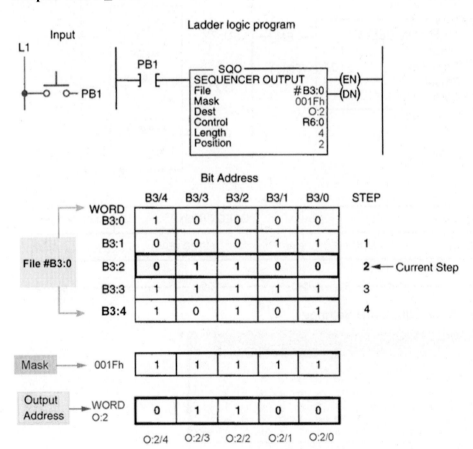

12-2(a) Implement the time-driven sequencer program shown, which is used for traffic light control at a four-way intersection. In this application, the control of traffic is accomplished using two Sequencer Output (SQO) instructions and a single Timer On Delay (TON) instruction. Use the Traffic Simulator screen and the addresses shown to simulate the program.

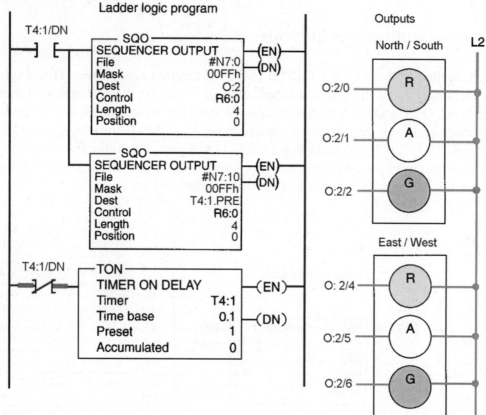

Ladder logic program

Outputs

T4:1/DN	SQO SEQUENCER OUTPUT	(EN) (DN)
	File #N7:0	
	Mask 00FFh	
	Dest O:2	
	Control R6:0	
	Length 4	
	Position 0	

	SQO SEQUENCER OUTPUT	(EN) (DN)
	File #N7:10	
	Mask 00FFh	
	Dest T4:1.PRE	
	Control R6:0	
	Length 4	
	Position 0	

T4:1/DN	TON TIMER ON DELAY	(EN)
	Timer T4:1	
	Time base 0.1	(DN)
	Preset 1	
	Accumulated 0	

North / South — L2

O:2/0 — R
O:2/1 — A
O:2/2 — G

East / West

O: 2/4 — R
O:2/5 — A
O:2/6 — G

Timing Chart

RED (N/S)		GREEN (N/S)	AMBER (N/S)
GREEN (E/W)	AMBER (E/W)	RED (E/W)	

<---------- 25 s ------->X---- 5 s --->X---------- 25 s -------->X--- 5 s -->

Sequencer File #N7:0 Light Cycle Settings

Integer Table

	15	14	13	12	11	10	9	8	7	6	5	4	3	2	1	0
N7:0/	o	o	o	o	o	o	o	o	o	o	o	o	o	o	o	o
N7:1/	o	o	o	o	o	o	o	o	o	1	o	o	o	o	o	1
N7:2/	o	o	o	o	o	o	o	o	o	1	o	o	o	o	o	1
N7:3/	o	o	o	o	o	o	o	o	o	o	o	1	o	1	o	o
N7:4/	o	o	o	o	o	o	o	o	o	o	o	1	o	o	1	o

Radix: Binary   Table: N7: Integer   Forces

Address: _____   Symbol: _____

Sequencer File #N7:10 Timer Settings

	Value
N7:10	0
N7:11	25
N7:12	5
N7:13	25
N7:14	5

Radix **Decimal** ▼   Table: **N7: Integer** ▼   Forces

Address _____   Symbol _____

Courtesy of TheLearningPit

12-2(b) Add the necessary programming to illuminate the appropriate Walk sign.

12-3 Implement the event-driven sequencer program shown. Use the I/O Simulator screen and the addresses and data file (#B3:0) shown to simulate the program.

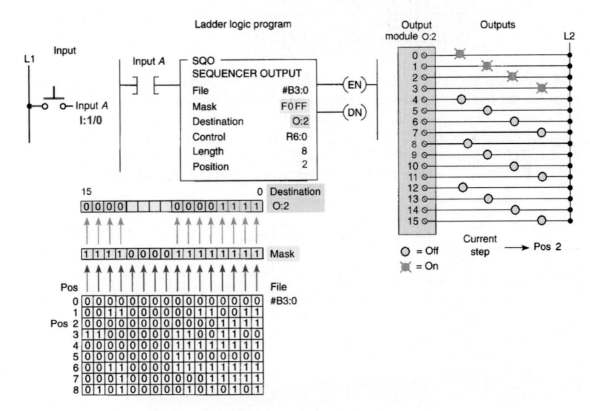

12-4 Implement the time-driven sequencer program shown using the same data (file #B3:0) that was used in the previous assignment (12-3). Demonstrate that the outputs are energized in the same sequence automatically at 3-second intervals. Use the I/O Simulator screen and the addresses shown to simulate the program.

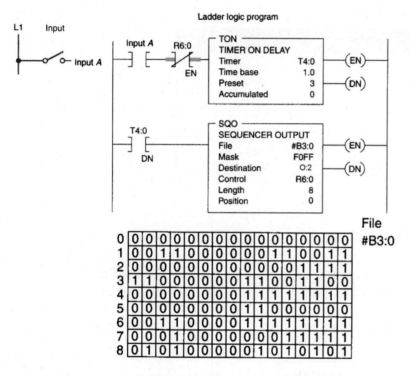

12-5 Implement the SEQUENCER COMPARE (SQC) program shown. Demonstrate that whenever the combination of opened and closed switches connected to I:1/12, I:1/13, I:1/14, and I:1/15 is equal to the combination of 1s and 0s on a step in the sequencer reference file, and the input I:1/0 is true, the PL1 output will become energized. Note how the mask (F000h) allows unused bits of the sequencer instruction to be used independently. Use the I/O Simulator screen and the addresses shown to simulate the program.

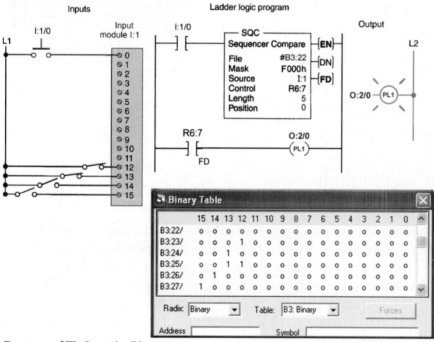

Courtesy of TheLearningPit

12-6 Implement the SEQUENCER LOAD (SQL) program shown. Demonstrate how the sequencer load instruction copies data from the source address to the file. Use the I/O Simulator screen and the addresses shown to simulate the program.

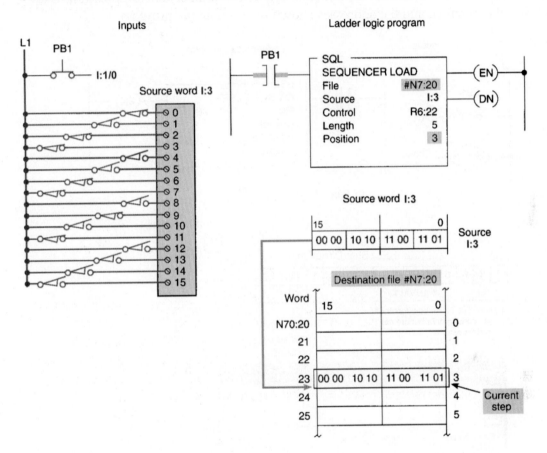

12-7 Implement the BIT SHIFT LEFT (BSL) program shown. Demonstrate that when a shift pulse is generated by a false-to-true transition of limit switch LS1, the enable bit is set and the data block is shifted to the left (to a higher bit number) one bit position. Use the I/O Simulator screen and the addresses shown to simulate the program.

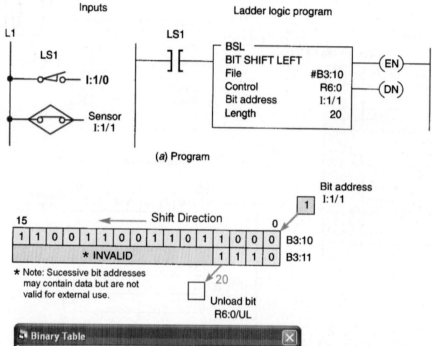

(a) Program

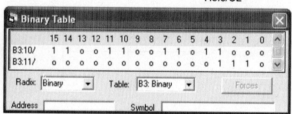

Data block array before shift pulse generated by LS1

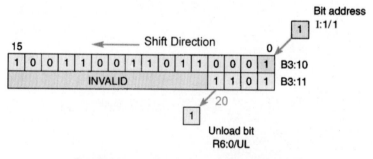

Data block array after shift pulse generated by LS1

12-8 Implement the BIT SHIFT RIGHT (BSR) program shown. Demonstrate that when a shift pulse is generated by a false-to-true transition of limit switch LS1, the enable bit is set and the data block is shifted to the right (from a higher address to a lower address) one bit position. Use the I/O Simulator screen and the addresses shown to simulate the program.

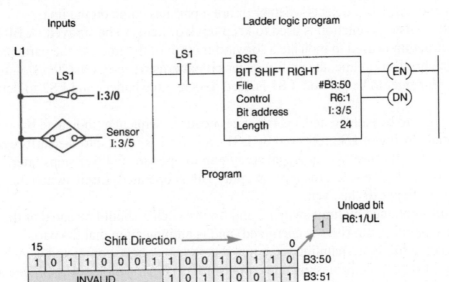

Data block array before shift pulse generated by LS1

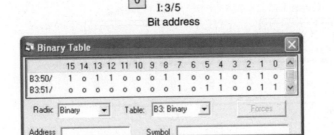

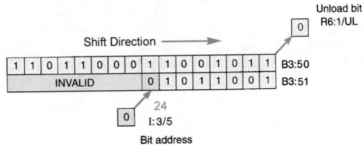

Data block array after shift pulse generated by LS1

12-9 Implement the shift register spray-painting application program shown. The sequence of operation is as follows:

➤ Each file bit location represents a station on the line, and the status of the bit indicates whether or not a part is present at that station.
➤ The bit address, I:1/2, detects whether or not a part has come on the line.
➤ The shift register's function is used to keep track of items to be sprayed. A Bit Shift Left instruction is used to indicate a forward motion of the line. As the parts pass along the production line, the shift register bit patterns represent the items on the conveyor hanger to be painted. LS1 is used to detect the hanger and LS2 to detect the part.
➤ When a part to be painted and a part hanger occur in sequence (indicated by a sequential closing of LS2 followed by LS1), a logic 1 is input into the shift register.
➤ The logic 1 will cause the undercoat spray gun to operate, and five steps later, when 1 occurs in the shift register, the topcoat spray gun is operated. Limit switch 3 counts the parts as they exit the oven.
➤ The counts obtained by limit switch 2 and limit switch 3 should be equal at the end of the spray-painting run (PL1 is energized) and is an indication that the parts commencing the spray-painting run equal the parts that have completed it. A logic 0 in the shift register indicates that the conveyor has no parts on it to be sprayed and, therefore, inhibits the operation of the spray guns.

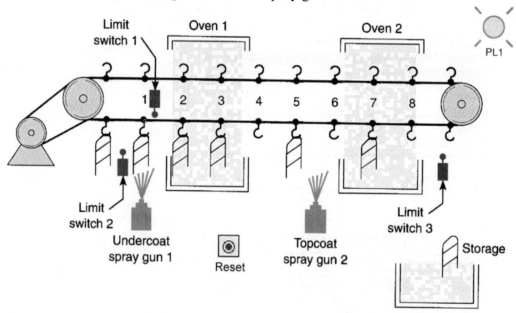

Use the I/O Simulator screen and the addresses shown to simulate the program.

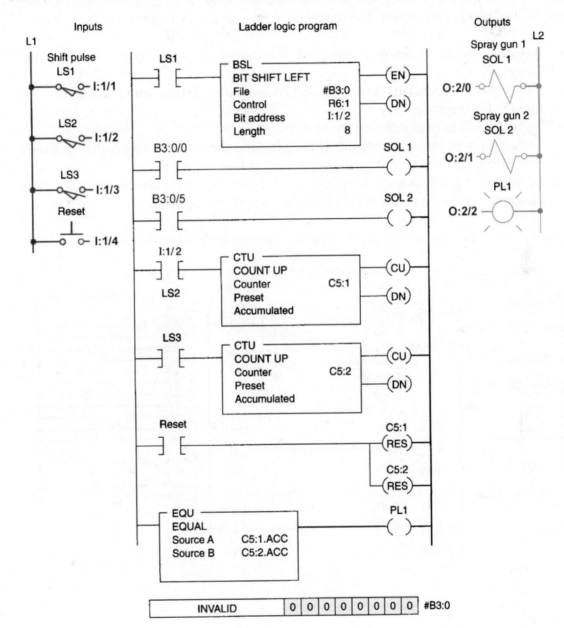

12-10 Implement the shift register program used to keep track of carriers flowing through the 16-station program shown. The sequence of operation is as follows:

➢ Proximity switch #1 senses a carrier, while proximity switch #2 senses a part on the carrier.

➢ Pilot lights connected to output module O:4 turn on as carriers with parts move through the machine.

➢ They turn off as empty carriers move through.

➢ Station #4 is an inspection station. If the part fails, the inspectors push PB1 as they remove the part from the system, which turns output O:4/4 off.

➢ Rework is added back into the system at station #6. When the operator puts a part on an empty carrier, he or she pushes PB2, turning output O:4/6 on.

Use the I/O Simulator screen and the addresses shown to simulate the program.

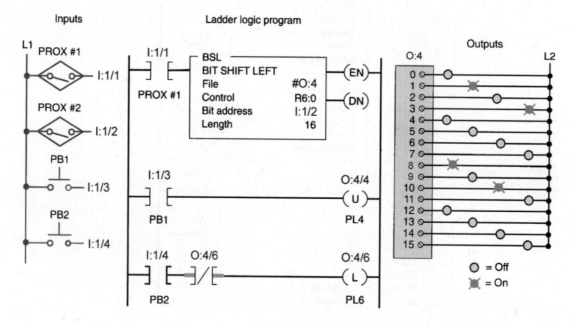

12-11 Implement the FIFO (first in, first out) instruction pair word shift register program shown. Both of the FIFO instructions are output instructions, and they are used as a pair. The FIFO LOAD (FFL) loads data from a source element; the FIFO UNLOAD (FFU) unloads instruction data from a file to a destination word. This program permits the stacking of data in a file. Two separate shift pulses are required: one to shift data into the file (LOAD) and one to shift data out of the file (UNLOAD).

Demonstrate how data are indexed in and out. Use the I/O Simulator screen and the addresses shown to simulate the program.

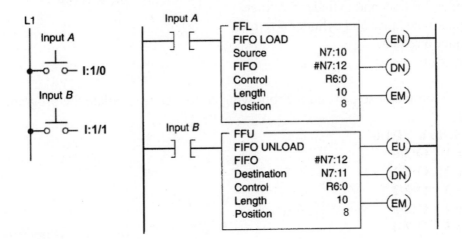

12-12 Using the sequencer output instruction, develop a time-driven program that will operate the cylinders in the desired sequence. The time between each step is to be 3 seconds. The desired sequence of operation is as follows:

➢ All cylinders to retract.
➢ Cylinder 1 advance.
➢ Cylinder 1 retract and cylinder 3 advance.
➢ Cylinder 2 advance and cylinder 5 advance.
➢ Cylinder 4 advance and cylinder 2 retract.
➢ Cylinder 3 retract and cylinder 5 retract.
➢ Cylinder 6 advance and cylinder 4 retract.
➢ Cylinder 6 retract.
➢ Sequence to repeat.

Use the I/O Simulator screen and the following addresses to simulate the program:

ON/OFF Switch _ I:1/0
Cylinder 1 _ O:2/0
Cylinder 2 _ O:2/1
Cylinder 3 _ O:2/2
Cylinder 4 _ O:2/3
Cylinder 5 _ O:2/4
Cylinder 6 _ O:2/5
TON _ T4:1

12-13 Using the sequencer output instruction, develop a program to implement an automatic car-wash process. The process is to be event-driven by the vehicle, which sequentially activates various limit switches (LS1 through LS6) as it is pulled by a conveyor chain through the car-wash bay. Design the program to operate the car wash in the following manner:

➢ The vehicle is connected to the conveyor chain and pulled inside the car-wash bay.
➢ LS1 turns the water input valve on.
➢ LS2 turns on the soap release valve, which mixes with the water input valve to provide a wash spray.
➢ LS3 shuts off the soap valve, and the water input valve remains on to rinse the vehicle.
➢ LS4 shuts off the water input valve and activates the hot wax valve, if selected.
➢ LS5 shuts off the hot wax valve and starts the air blower motor.
➢ LS6 shuts off the air blower. The vehicle exits the car wash.

Use the I/O Simulator screen and the following addresses to simulate the program:

ON/OFF Switch _ I:1/0
LS1 _ I:1/1
LS2 _ I:1/2
LS3 _ I:1/3
LS4 _ I:1/4
LS5 _ I:1/5
LS6 _ I:1/6
Hot Wax Switch _ I:1/7
Water Input Valve _ O:2/1
Soap Release Valve _ O:2/2
Hot Wax Valve _ O:2/3
Air-Blower Motor _ O:2/4

12-14 A product moves continuously down an assembly line that has four stations, as shown. The product enters the inspection zone, where the proximity switch senses its presence. The inspector examines it and activates a reject switch if the product fails inspection. If the product is defective, reject status lights come on at stations 1, 2, and 3 to tell the assembler to ignore the part. When a defective part reaches station 4, a diverter gate is activated to direct that part to a reject bin. Develop a PLC program that uses the Bit Shift Left instruction to implement this process.

Use the I/O Simulator screen and the following addresses to simulate the program:

Reject Switch (NO) _ I:1/0
Proximity Switch (NC) _ I:1/1
Reject Light 1 _ O:2/0
Reject Light 2 _ O:2/2
Reject Light 3 _ O:2/4
Diverter Gate _ O:2/6

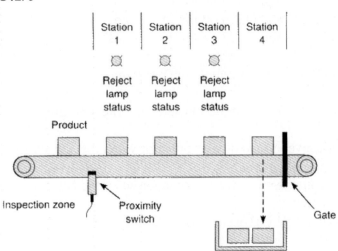

12-15(a) Using the sequencer output (SQO) instruction, design a program to turn on outputs according to the eight steps of the matrix-style chart shown. Employ a pushbutton to step through the table. Use the I/O Simulator screen and the following addresses to simulate the program:

Pushbutton (NO) _ I:1/0
Outputs 0 through 7 _ Output Word O:2

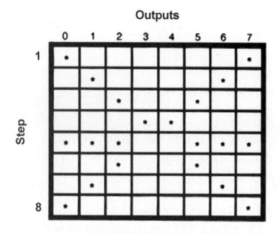

12-15(b) Modify the original program to operate continuously by using a recycling timer's done bit to trigger a step in the sequence at 1-second intervals.

12-16 Create a file of timer preset values, and create a sequencer output file. The file of timers will work in conjunction with the sequencer so that the outputs remain on for a period of time derived from the timer table. The sequencer should remain in a particular step until the timer times out and then proceeds to the next step as follows:

➢ Step 1. Outputs O:2/0 and O:2/4 on for 5 seconds
➢ Step 2. Outputs O:2/1 and O:2/5 on for 10 seconds
➢ Step 3. Outputs O:2/2 and O:2/6 on for 15 seconds
➢ Step 4. Outputs O:2/3 and O:2/7 on for 20 seconds

Use the I/O Simulator screen to simulate the program.

12-17(a) Using the Bottle Line simulation screen, write a program that

➢ Starts the main conveyor and divert conveyor when the NO start pushbutton (**I:1/1**) is momentarily pressed.
➢ Stops the main conveyor and divert conveyor when the NC stop pushbutton (**I:1/0**) is momentarily pressed.
➢ Uses sensor switch LS1 (**I:1/6**) to detect all bottles.

- ➢ Uses sensor switch LS3 **(I:1/8)** to detect broken bottles.
- ➢ Uses the Bit Shift Left instruction to automatically energize the divert gate **(O:2/5),** thus transferring all broken bottles to the divert conveyor.

12-17(b) Change the original program so that the Bit Shift Left instruction automatically energizes the divert gate to transfer all large bottles to the divert conveyor. Use sensor switch LS2 **(I:1/7)** to detect large bottles.

12-17(c) Modify the program to operate as follows:

- ➢ Use one Bit Shift Left instruction to automatically energize the scrap gate **(O:2/4)** to transfer all broken bottles to the grinder **(O:2/3).**
- ➢ Use a second Bit Shift Left instruction to automatically energize the divert gate to transfer all large bottles to the divert conveyor.
- ➢ Have the grinder motor operate any time the main conveyor motor is operating.
- ➢ Automatically stop the main conveyor motor whenever the scrap conveyor (Enter) NO pushbutton **(I:1/5)** is closed. With the main conveyor stopped, allow the operator to manually operate the scrap conveyor motor **(O:2/1)** using this pushbutton and utilizing the box in place sensor switch LS10 **(I:1/15)** to move an empty scrap box into position.

CHAPTER **13**

PLC Installation Practices, Editing, and Troubleshooting

LogixPro Programming Assignments

13-1 With reference to the ladder logic program shown, view the input and output image tables to compare the logic status of the hardwired limit switch and pilot light with their logic state. Use the I/O Simulator screen and the following addresses to simulate the program:

Limit Switch _ I:1/3
PL1 _ O:2/3

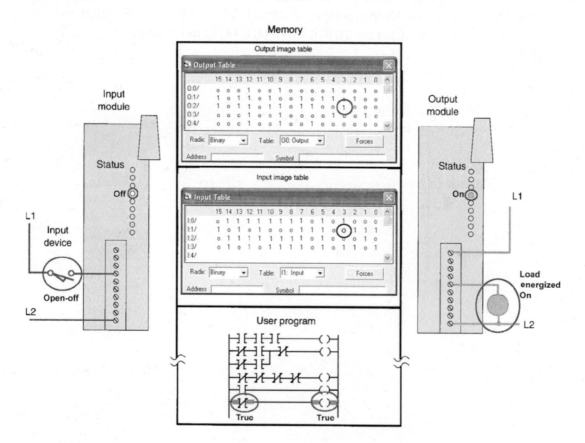

Courtesy of TheLearningPit

13-2 With reference to the ladder logic program shown, add instructions to modify the program to ensure that Pump 2 does not run while Pump 1 is running. If this condition occurs, the program should suspend operation. Use the I/O Simulator screen and the following addresses to simulate the program:

Switch 1 _ I:1/0
Switch 2 _ I:1/1
Pump 1 _ O:2/0
PL1 _ O:2/1
Pump 2 _ O:2/2

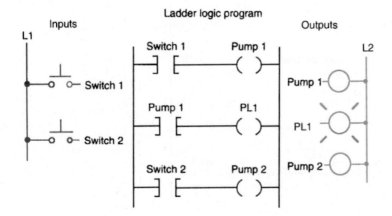

13-3 The PLC program shown is supposed to execute to sequentially turn PL1 off for 5 seconds and on for 10 seconds whenever input A is closed. Troubleshoot the circuit, and identify what needs to be changed to have it operate properly. Use the I/O Simulator screen and the following addresses to simulate the program:

Input A _ I:1/0
PL1 _ O:2/0
Internal Relay _ B3:1/0
Timers _ T4:2 & T4:3

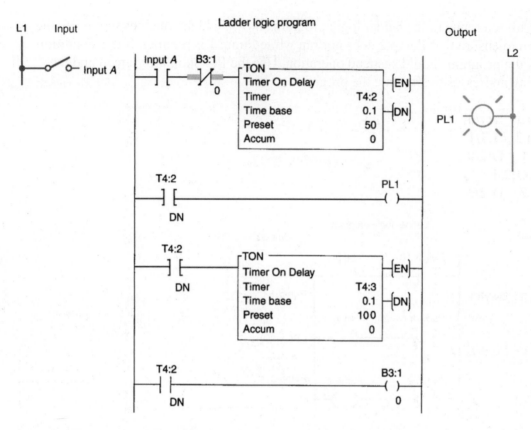

13-4 Construct a simulated PLC program for the program editing and control exercise shown. Use the I/O Simulator screen and the following addresses to simulate the program:

Input A _ I:1/0	**Output I _ O:2/0**
Input B _ I:1/1	**Output J _ O:2/1**
Input C _ I:1/2	**Output K _ O:2/2**
Input D _ I:1/3	**Output L _ O:2/3**
Input E _ I:1/4	**Output M _ O:2/4**
Input F _ I:1/5	
Input G _ I:1/6	
Input H _ I:1/7	

➤ Enter the original program into the PLC, and prove its operation.
➤ Enter an additional rung following rung 4. This new rung is to examine the status bit of output **I** for an OFF condition and energize output **M** when the rung condition is true.
➤ Remove the instruction from rung No. 2 that examines address **E** for an Examine If Closed condition.

➢ Place the Examine If Closed instruction **(E)** that was removed in the previous step into rung 7 of the original program. Insert this Examine If Closed instruction in parallel with the existing Examine If Closed instruction **H**.

➢ While in the Run mode, change the preset value of the counter from 10 to 25.

➢ Remove rung 5 of the original program.

➢ While in the Run mode, force output **K** to an ON condition.

➢ While in the Run mode, force output **J** to an OFF condition.

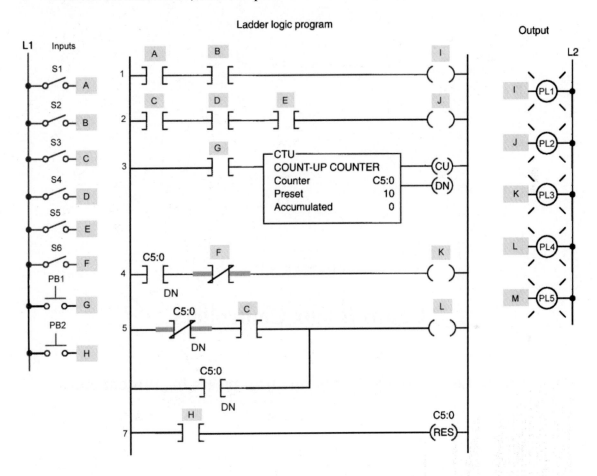

CHAPTER **14**

Process Control, Network Systems, and SCADA

There are no LogixPro programming assignments for this chapter.

CHAPTER **15**

ControlLogix Controllers

There are no LogixPro programming assignments for this chapter.